D0756041

WITHDRAWN

The Fundamentals of Quality Management

Dennis F. Kehoe
Department of Industrial Studies
University of Liverpool, UK

SPRINGER-SCIENCE+BUSINESS MEDIA, B.V.

First edition 1996

Originally published by Chapman & Hall 1996

Typeset in by 10/12 pt. Times by Saxon Graphics Ltd, Derby

ISBN 978-0-412-62690-6 ISBN 978-94-011-0545-3 (eBook)
DOI 10.1007/978-94-011-0545-3

A catalogue record for this book is available from the British Library
Library of Congress Catalog Card Number: 95-70859

♾ Printed on permanent acid-free text paper, manufactured in accordance with ANSI/NISO Z39.48–1992 and ANSI/NISO Z39.48–1984 (Permanence of Paper).

The Fundamentals of
Quality Management

To my children, to Sarah and to Andrew

Contents

Preface

This book has been written to provide both students and industrial managers with a comprehensive description of the tools and techniques of Quality Management and also to provide a framework for understanding Quality Development.

Central to the theme of this book is the idea that quality management is a developmental process which requires an understanding of the techniques, the people and the systems issues. The aims of quality development are to produce greater organizational consistency, to improve customer satisfaction and to reduce the business process costs. In order to achieve these aims, managers are required to have an understanding of both the underlying theories and the methodologies for implementation. The aim of this book is to provide a coherent description of both the theoretical and implementation aspects of quality management.

Since the halcyon days of the quality 'revolution' of the 1970s and 1980s, many organizations have realized that quality development represents an enormous management challenge. This challenge for continuous improvement requires the continuous development of systems, of techniques and of people. Like most serious business strategies, competitive improvement through quality development can only be achieved if the organization understands not only what the various quality 'options' are but also when a particular technique or approach is applicable. Quality development has no single blueprint but requires a learning organization which understands key concepts and methods of implementation.

The inspiration for this book came from the teaching of a great many students, researchers and industrialists and from experiencing their problems of understanding and application. Providing students with a framework for understanding the role of the various techniques of quality management and by proposing for industrial managers a range of implementation methodologies have been the 'motivators' for producing this text.

Many people have assisted directly and indirectly in the preparation of this book. My special thanks go to Christine Williams for her tireless work in preparing the manuscript and to Harry Evison for his painstaking proof-reading and valuable suggestions for improvement. Read this book with a critical eye and form your own views on the relevance and applicability of the techniques described. For most organizations and the individuals who work within them the only choice in the future is **how** to apply the techniques of quality management, **not** whether to.

Dennis F. Kehoe
Liverpool, January 1995

1 Introduction

1.1 The definition of quality and the role of quality management

The aim of this book is to provide the reader with an understanding of the fundamental concepts of managing quality. The starting point for most individuals – and indeed most organizations – is to understand the meaning of the term 'quality'. This may seem a rather elementary requirement as the word 'quality' is recognized by most people and in general speech is used to describe excellence, value, reliability or goodness. The Oxford English dictionary defines quality as:

(Noun) Degree of excellence, relative nature ...

In a business context, however, the concept of quality needs to be more precisely understood and to be clearly interpreted by everyone in the organization. In essence quality can be understood as:

Meeting the customer's expectations.

In order to manage quality, organizations need therefore to have mechanisms in place both to establish what the customer expects (or requires) and to confirm that these expectations have been met.

The effective management of quality requires the tools and techniques described throughout this book to be applied systematically and within an overall framework of development. The management challenge to understand and implement these fundamental concepts is significant. What therefore are the benefits? Why, in other words, is quality management important?

The simple answer is competition. As international markets move towards the millennium then competition will continue to increase. The international markets for many products are now very mature – the products or services offered by different companies from different countries are basically the same. As a result the basis of competition is increasingly the quality of the product or service. There is no conflict between quality and price as more and more organizations realize that in the long term quality goods and service actually cost less. A nineteenth-century economist once observed that 'if you spend too much on your goods, all you lose is a little money; if you spend too little then you can lose everything.'

1.2 The development of quality management

The issue of quality of goods or services is not new. Throughout history society has demanded that providers of goods or services should meet their obligations. As long ago as 1700 BC King Hammurabi of Babylon introduced the concept of product quality and liability into the building industry of the time by declaring:

> ... if a building falls into pieces and the owner is killed then the builder shall also be put to death. If the owners' children are killed then the builders' children shall also be put to death.

During the Middle Ages many of the guilds of craftsmen were established to guarantee the quality of workmanship and to define the standards to be expected by the purchaser. During the Industrial Revolution, many of the technological advances, such as the development of the steam engine, were made possible through developments in metrology and the standardization of engineering components such as screw threads.

The advent of mass production during the twentieth century increased the demands upon the control of product quality. During the 1940s and 1950s the techniques of quality control became an increasingly important aspect of business management as organizations sought to gain competitive advantage and reduced costs through the inspection of product quality. The success of Japanese manufacturers during the 1960s and 1970s changed the emphasis from a quality control approach to a quality assurance approach requiring more of the business functions to be involved in the management of quality and requiring longer implementation timescales. Finally the fierce international competition for goods and services during the 1980s and 1990s has led to a 'total' approach to quality management whereby everyone in the organization is involved in developing an improvement and prevention orientation which focuses upon the customer through teamwork. The timescales for implementing total quality are even longer and the relationship between the involvement of people and the implementation timescales for the historical development of quality management is illustrated in Figure 1.1.

1.3 A framework for quality management

The framework proposed throughout this book is that quality improvement is a developmental process. The fundamental tools and techniques of quality management can therefore be 'mapped' onto this developmental framework to provide a better understanding of the role of each of the techniques and the context of its implementation.

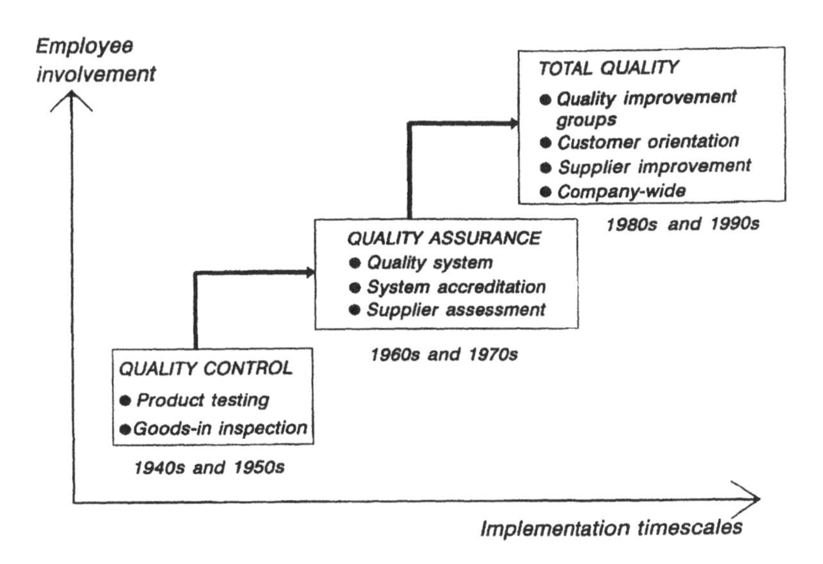

Figure 1.1 The development of quality management approaches.

The stages of development that most organizations progress through are successively:

- a systems orientation;
- an improvement orientation;
- a prevention orientation.

This development in the way companies manage quality is 'encompassing' in the sense that the developments made as the quality systems are implemented need to be retained and built upon as the organization moves towards improvement. Prevention in turn requires the improvement activities to be augmented with a greater emphasis upon designing quality into the product or service or process rather than improving existing ones. This developmental framework illustrated in Figure 1.2 is used throughout the book both to position the quality tools, techniques and methodologies described and also to provide an understanding of the effects of the various activities.

Chapter 2 describes the basic concepts of a systems approach to quality and the role of ISO 9000 in promoting an international standard for assessing the quality management systems employed in organizations. The implementation of quality systems represents a fundamental foundation for quality development and the promotion of a company-wide quality perspective.

The enablers for quality improvement are presented in Chapter 3, which examines the measurement of the cost of quality, and Chapter 4, which describes the involvement and motivation of people for quality and the role

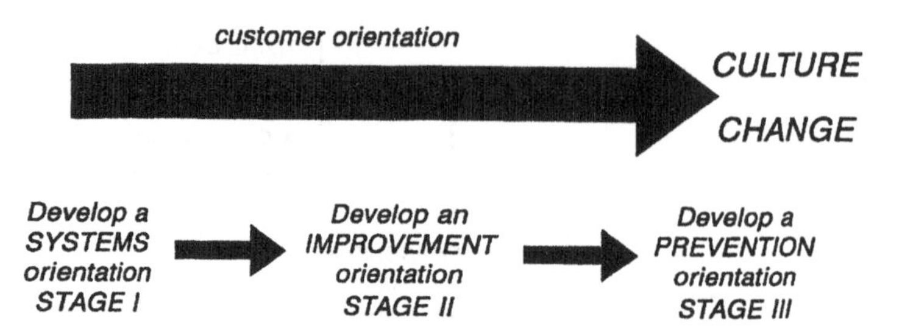

Figure 1.2 The quality management development framework.

of teamwork and leadership. The coordination of quality improvement activities is outlined in Chapter 5 together with the cultural changes required to create total quality.

The fundamental techniques of quality management are covered in Chapters 6, 7 and 8. Chapter 6 looks at the application of acceptance sampling and the role of sampling in each of the stages of quality development. The techniques of statistical process control (SPC) and the implementation of process improvements through a reduction in variation are described in Chapter 7 which also outlines the fundamental methodology required to implement SPC. Chapter 8 examines both the basic and the advanced problem-solving tools required to facilitate team-based improvement.

Finally Chapters 9 and 10 describe the prevention-orientated methods associated with a developed approach to quality management. Product and systems reliability is covered in Chapter 9 which outlines the analysis techniques required to predict product or process failures and to design more robust systems. Chapter 10 outlines advanced quality planning techniques including methodologies such as quality function deployment for improved product design and experimental design for establishing processes less susceptible to variation.

1.4 Systems, culture, tools and customers

For most organizations the challenge of quality development requires progress in each of the three basic dimensions of quality management:

- people;
- systems;
- techniques.

This book describes the developments required in each of these aspects of quality in order to provide improved customer satisfaction and enhanced business performance. It is through the systematic integration of all three dimensions that organizations achieve the breakthrough in operational performance and customer service associated with total quality. Progress, however, requires both an understanding of the fundamental principles of quality management and also an appreciation of the implementation approaches necessary for practical success.

Quality development has a potentially significant impact upon all three elements of business – customers, shareholders and employees. Improved customer service enhances customer loyalty and generates increased revenues. More effective internal operations reduce quality costs and hence improve business performance. Finally a quality culture in which people are empowered creates increased job satisfaction and therefore a more motivated workforce.

The dimensions of quality management are illustrated in Figure 1.3.

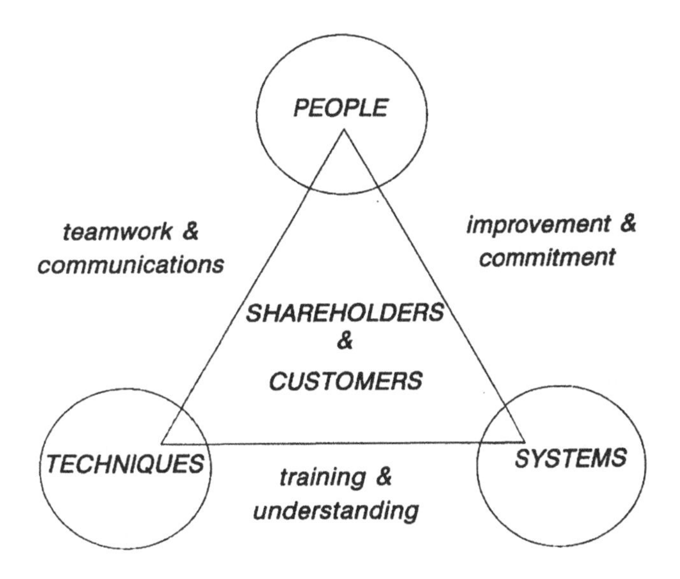

Figure 1.3 The dimensions of quality management.

The benefits from understanding the fundamentals of quality management are significant for all types of organization and represent a major contributor to business performance in the years ahead.

2 Quality systems and ISO 9000

2.1 Quality system standards

2.1.1 BACKGROUND TO QUALITY SYSTEM STANDARDS

In order to fully appreciate the extremely important business developments in relation to quality system standards, one first has to understand the meaning of the term 'quality'. A number of informative definitions have emerged:

- 'the features and characteristics of a product or service which bear upon its ability to satisfy a stated or implied need' (BS 4778, ISO 8402);
- 'conformance to specification';
- 'fitness for purpose';
- 'meeting customers' requirements, and exceeding their expectations';
- 'doing things right first time'.

The essence of this approach to quality is to provide company operating systems which promote conformance to specification. Therefore:

- specifications (or requirements) need to be clearly defined;
- a 'systems' approach is required involving all aspects of the business;
- responsibilities, processes, methods and materials all need to be prescribed.

It is from this holistic view of quality which developed during the late 1970s that the concept of a **Quality System** began to gain widespread industrial acceptance. The limitations of product standards, the move towards a quality assurance rather than quality control philosophy and the ever increasing pressure to provide better and better quality products and services led to the adoption of the concept of a quality system which could be defined as:

> The organizational structure, responsibilities, procedures, processes and resources for implementing quality management.

The emergence of this 'systems based' approach to the management of

Quality led to the need for a generalized standard for assessing such quality systems which provided:

- a general framework or agenda for assessing a company's quality system;
- a structure which is applicable to all organizations, from manufacturing to service, from large to small;
- an independently verifiable, internationally accepted systems checklist.

In terms of what constituted a quality system the approach adopted has been that this covers all business functions with the exception of finance (the argument here being that financial systems have alternative, often legal audit requirements). The main elements of a quality system are shown in Figure 2.1.

Whilst the main sector for the implementation of quality systems has been manufacturing industry, the concept can also be applied to service companies, in which case the term 'manufacturing' is taken to mean transaction processing.

2.1.2 DEVELOPMENT OF QUALITY SYSTEM STANDARDS

The late 1970s saw increasing international pressure for improvements in the management of quality and in particular for standards for assessing the way in which quality was managed within a supplier company. These pressures included:

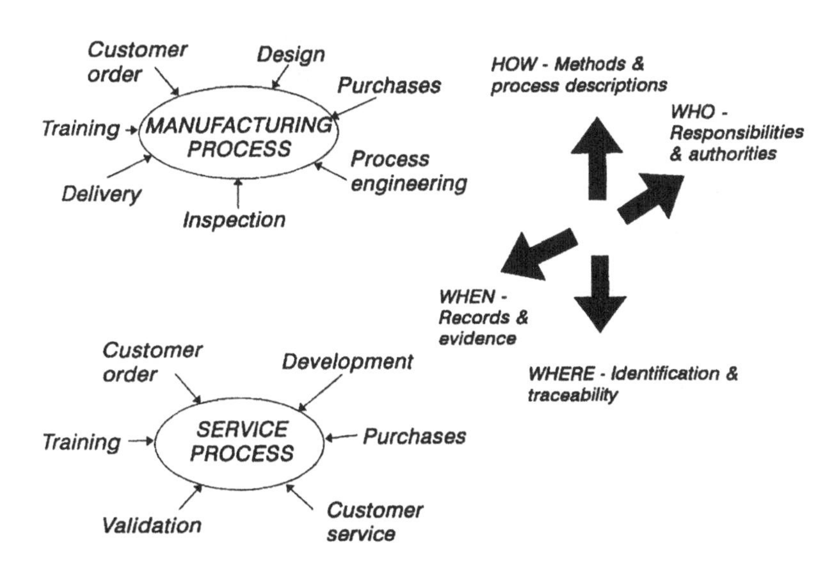

Figure 2.1 Typical elements and dimensions of a quality system.

- increased international competition, particularly from Japanese manufacturers;
- the proliferation of multiple assessments whereby supplier companies would be assessed by many different customers;
- increasing product liability concerns and the limitations of product standards in providing supplier protection.

The developments in the UK during this period were typical of international trends in the adoption of quality system standards:

- 1974 – the publication of BS 5173 *Guide to the Operation and Evaluation of Quality Systems*;
- 1979 – the publication of BS 5750 by the British Standards Institution which is written by industry for industry as the national third-party assessment standard;
- 1982 – the UK government publishes a White Paper in which it is proposed to specify conformance to BS 5750 as a requirement in government contracts;
- 1983 – the International Standards Organization (ISO) recognizes the widespread interest in quality systems standards and begins work on an international version based upon the UK national standard;
- 1985 – the launch of the National Accreditation Council for Certification Bodies (NACCB) by the Secretary of State for Trade and Industry to establish competencies and impartiality for bodies providing certification to BS 5750;
- 1987 – the Consumer Protection Act provides an incentive to businesses to ensure the quality of their products or services in order to limit liability.

All these developments and plans culminated in the publication in 1987 of a range of common national and international quality system standards entitled *Quality Systems – Model for Quality Assurance*.

- 1992 – the European Community moves towards a common trading market which highlights the need for a Community-wide supplier assessment standard;
- 1994 – revision of international quality system standards.

These standards, which were identical in content, included:

- BS 5750: 1987 Parts 0, 1, 2 and 3 (from 1994 entitled BS EN ISO 9000 series);
- ISO 9000 Series 9001, 9002, 9003 and 9004;
- EN 29000 Series 29001, 29002, 29003 and 29004.

Although these standards are issued and updated by the appropriate standards organizations (British Standards Institution, International Standards Organization, European Committee for Standardization), the accreditation to the standard can be awarded by any one of a number of organizations. These include bodies approved by the NACCB, for example:

- British Standards Institution, Lloyds, SGS Yarsleys, etc.,

and national standards bodies, for example:

- Singapore Institute for Standards and Industrial Research, Standards and Industrial Research Institute of Malaysia.

The national – and indeed international – accreditation to BS 5759/ISO 9000/EN 29000 has become a very standardized process and there exists only marginal differences between the methodologies adopted by the different certification bodies. The great strength of these standards has been their universal acceptance and application using a common format; together they represent the single most important international development in quality management since the Second World War.

2.1.3 TYPES OF QUALITY SYSTEM STANDARDS

Quality system standards specify a set of minimum quality assurance requirements and are primarily used for:

- guiding supplier companies who are introducing quality assurance and helping in the structure of the quality system design;
- providing a basis for assessing or evaluating a supplier's (or potential supplier's) quality system;
- establishing a contractual basis for quality system requirements.

Quality system standards are to be contrasted therefore with product standards which define the product requirements for a particular item or group of items (for example, electrical devices, safety equipment, etc.). Quality system standards are concerned with the way in which the company is organized, and operate and are independent of the product or service provided. Basically there are two types of standard.

Industry-related standards are used by (usually large) purchasing bodies, for example the defence, aerospace or nuclear industries. A typical example of an industry-based standard is the Ford Motor Company's Q101 Worldwide Supplier Quality System Standard shown in Figure 2.2.

The basis of the Q101 is an assessment of each of the 20 elements. These are each marked out of a potential score of ten with a minimum total score of 140 required with no individual element scoring less than five marks.

Another widely adopted industry-specific quality system standard is the AQAP series used by the defence industries:

- AQAP 1–3 NATO Requirements for an Industrial Quality Control System;

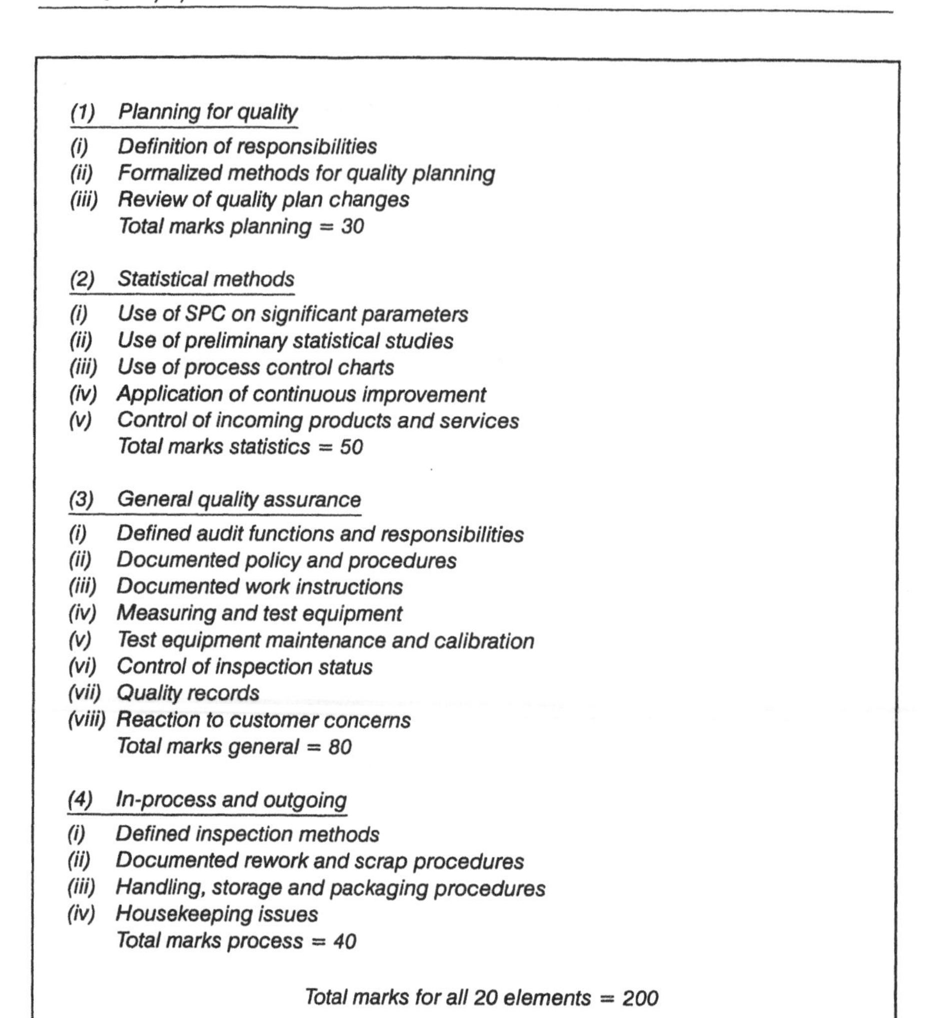

(1) Planning for quality

(i) Definition of responsibilities
(ii) Formalized methods for quality planning
(iii) Review of quality plan changes
Total marks planning = 30

(2) Statistical methods

(i) Use of SPC on significant parameters
(ii) Use of preliminary statistical studies
(iii) Use of process control charts
(iv) Application of continuous improvement
(v) Control of incoming products and services
Total marks statistics = 50

(3) General quality assurance

(i) Defined audit functions and responsibilities
(ii) Documented policy and procedures
(iii) Documented work instructions
(iv) Measuring and test equipment
(v) Test equipment maintenance and calibration
(vi) Control of inspection status
(vii) Quality records
(viii) Reaction to customer concerns
Total marks general = 80

(4) In-process and outgoing

(i) Defined inspection methods
(ii) Documented rework and scrap procedures
(iii) Handling, storage and packaging procedures
(iv) Housekeeping issues
Total marks process = 40

Total marks for all 20 elements = 200

Figure 2.2 Ford Motor Co. Q101 System Standard.

- AQAP 4–2 NATO Inspection System Requirements for Industry;
- AQAP 9–2 NATO Basic Inspection Requirements for Industry.

General standards which are those published by national standards bodies for general use throughout industry and commerce. In recent years there has been an international convergence of these standards, as illustrated by the comparison of national quality system standards shown in Table 2.1.

The main advantages for businesses in the publication of these quality system standards are as follows:

- They provide a common framework for assessing the management of quality and this removes the variability of traditional customer assessments.

Table 2.1 Comparison of national quality system standards

National standards body	ISO 9001 – Design, Manufacture and Test	ISO 9002 – Manufacture and Test	ISO 9003 – Test
United Kingdom	BS 5750: Part1	BS 5750: Part2	BS 5750: Part3
Australia	AS 3901	AS 3902	AS 3903
Canada	CSAZ 299.1-85	CSAZ 299.2-85	CSAZ 299.4-85
India	IS 10201: Part4	IS 10201: Part5	IS 10201: Part6
Ireland	IS 300: Part1	IS 300: Part2	IS 300: Part3
Singapore	SS 308: Part1	SS 308: Part2	SS 308: Part3
Europe	EN 29001	EN 29002	EN 29003
USA	ANSI/ASQC Q91	ANSI/ASQC Q92	ANSI/ASQC Q93

- They provide for an international 'currency' in terms of the evaluation of a company's quality management which is important in export markets.
- They provide an internal focus for quality development which, because it is externally assessed, assists in the management of change process.

The main disadvantages of these universally applied quality system evaluation standards are, however, as follows:

- They represent a 'minimum' standard and therefore do not provide for long-term competitive advantage.
- They are mainly directed towards manufacturing or manufacturing engineering industries and therefore require interpretation for other business sectors.
- They are written in general terms and therefore open to interpretation.

2.1.4 THE BENEFITS OF ISO 9000 ACCREDITATION

The international quality system standard ISO 9000 has been widely applied across a range of manufacturing industries from engineering to textiles, from chemicals to food, and increasingly in non-manufacturing sectors including banks, insurance companies, hotels and colleges of further education. The benefits of gaining accreditation to ISO 9000 can be summarized as follows:

- National and international recognition of the quality systems employed at the company is increasingly seen as an essential prerequisite to international trade. Many countries viewed the developments in ISO 9000 in the late 1980s as part of a 'Fortress Europe' policy to prevent imports to the European Community. The Standard has, however, been adopted throughout the world as the basis for good management practice for quality assurance. The typical international distribution of ISO 9000 accreditation by 1994 is shown in Table 2.2.

Table 2.2 International distribution of ISO 9000 accreditation

Country	Approximate number of assessed companies (1994)
United Kingdom	28 000
Australia	2 700
USA	1 600
Germany	1 600
France	1 500
Netherlands	1 500
Singapore	500
Japan	400

In many sectors of industry it has become a minimum supplier requirement so that companies which do not possess ISO 9000 accreditation are not even asked to tender. The international chemicals group ICI has over 100 group companies approved to ISO 9000/BS 5750. In a recent survey (shown in Figure 2.3) most companies implementing ISO 9000 had experienced an increase in turnover due to the opening of new markets and improvements in companies' quality profile.

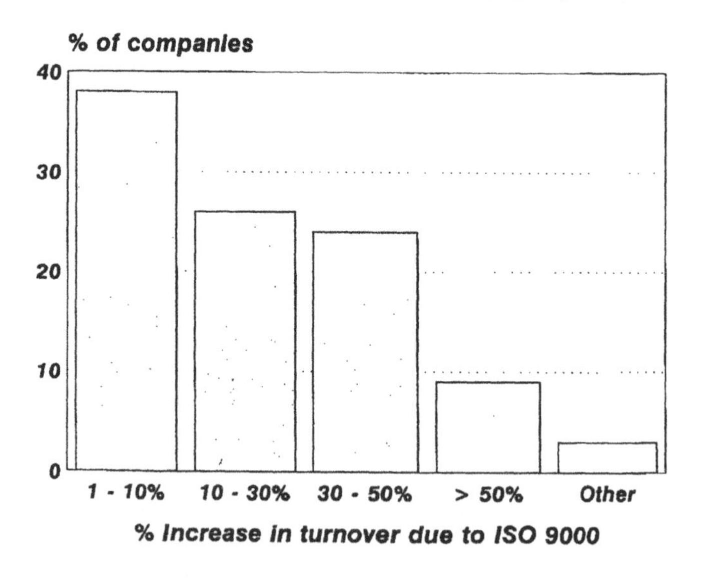

Figure 2.3 Percentage increase in turnover due to ISO 9000.

- As more and more organizations have adopted ISO 9000 as the supplier standard this has reduced the need for multiple assessment. This provides national benefit in terms of reducing the overall expenditure on quality assessments and also reduces the number of assessments which have to be hosted by individual companies. Individual companies can also be as certain as the assessor as to what the agenda for assessment will be. The requirement of the ISO 9000 standard for an approved company to in turn have assessed its own suppliers (see section 2.2.3) has led to a 'mushrooming' effect whereby companies that have gained accreditation begin to impose ISO 9000 on their suppliers. The survey of ISO 9000 approved companies illustrated this effect as shown in Figure 2.4.

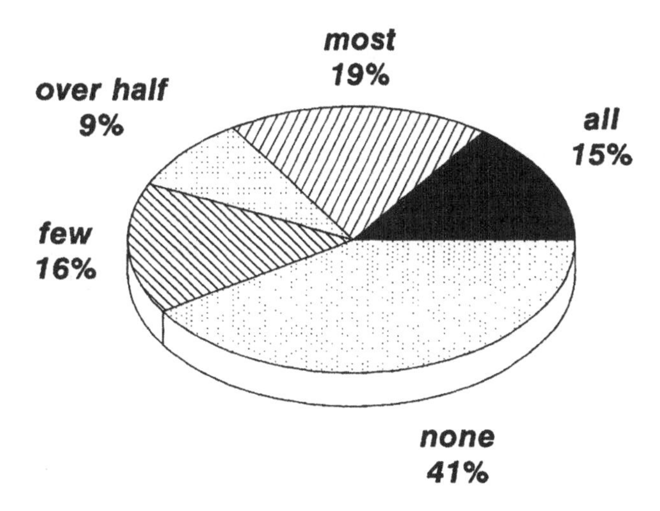

Figure 2.4 Percentage requirement for suppliers to conform to ISO 9000.

- ISO 9000 represents the basis for most companies' quality development by providing a foundation for future quality improvements. The Standard lends itself to implementation even before the organizational culture change necessary for total quality management (see Chapter 5) is in place.
- Many companies have become more and more concerned with the potentially very damaging implications of product liability litigation. The enormous cost of industrial health and safety tragedies has led insurance agents to seek assurances as to the way in which client companies manage product quality. ISO 9000 systems are seen to provide documented evidence of conformance to contractual specifications and therefore lawyers and insurers have increasingly pressed for clients working in safety-critical areas to be ISO 9000 approved.
- ISO 9000 has increasingly become the basis of a range of product and process standards. Product approval in the UK now requires the manufacturer to have ISO 9000/BS 5750 accreditation, and in areas such as

hazard and critical control point (HACCP) analysis in the food industry ISO 9000 is seen as an enabling system.

- On a personal level, individual employees are seeing a number of benefits in working for ISO 9000 approved organizations. These include the availability of well-defined methods and procedures which help the induction of new employees and a readily transferable appreciation of quality system requirements. This applies to all types of employees, from designers to production engineers, from purchasing managers to clerical staff, as well as the more obvious quality specialists.

Many of the commentators on quality management, particularly those in the USA and Japan, have expressed concern that the European-dominated ISO 9000 approach does not deliver 'world-class' manufacturing performance and 'gurus' such as Juran have suggested that quality system standards will not deliver continuous quality improvement. The point that is often missed, however, is that ISO 9000 has its major impact on organizations who are at an early stage of quality development and in particular the Standard has had a significant affect on small to medium sized companies who are not necessarily seeking world-class status. Indeed, for companies who are very mature in their management of quality then ISO 9000 may not offer significant developmental benefits.

2.1.5 ENVIRONMENTAL MANAGEMENT SYSTEMS AND BS 7750

A description of the development of quality management systems would be incomplete without reference to the growing and increasingly important development of environmental management system standards. In the UK the British Standard BS 7750 was published in 1992 entitled *Environmental Management Systems*. As with BS 5750 this environmental standard is seen as a forerunner to a European and international standard in the manner in which BS 5750 led to the publication of ISO 9000.

The BS 7750 standard has attempted to build upon the success, in terms of widespread acceptance, of the ISO 9000 series and has been deliberately aligned to the philosophy and structure of the quality system standard. To some extent this aligning has been reciprocated in the 1994 version of ISO 9000 which places increased emphasis upon health and safety environmental issues. When published in 1992, the BS 7750 standard was seen as a natural progression for companies which had achieved ISO 9000 and was written by a wide range of organizations and environmental bodies with an implemented quality system standard as a preferred starting point.

As with ISO 9000, the basis for BS 7750 is that the company should have a stated policy (on environmental management) and should be able to objectively demonstrate compliance (normally through audit). Since 1992 the BS 7750 standard has gained increasing acceptance in the UK, particularly in sectors of industry such as food and drinks manufacture, consumer goods and in power generation. A wide range of environmental elements are covered in the standard including air emissions, water resources and supplies, waste (including toxic waste) management, noise, radiation, environmental impact assessment and public safety.

The scope of BS 7750 covers the design, implementation, operation and audit of an environmental management system. The standard itself does not, however, specify environmental performance criteria, rather in the manner in which ISO 9000 described below in section 2.2 does not specify product quality criteria. This does not mean, however, that BS 7750 allows the organization to propose its own environmental performance criteria (for example, emission levels or noise standards) in the way acceptance criteria for product or service quality are established between the customer and supplier in ISO 9000. Instead BS 7750 requires the organization to comply with established environmental standards (for example, local authority pollution controls or building regulations).

The implementation of BS 7750 essentially requires the development of an environmental management system documented using the three elements shown in Figure 2.5.

The **environmental manual** is equivalent to the quality manual (described below in section 2.3) and provides an overview of the organization's environmental management system, including a description of the policy on the BS 7750 requirements together with a definition of responsibilities and an identification of the documentation used. The **register of regulations**, as the name suggests, is a catalogue of relevant environmental regulations (industry, local, national or international) which form the

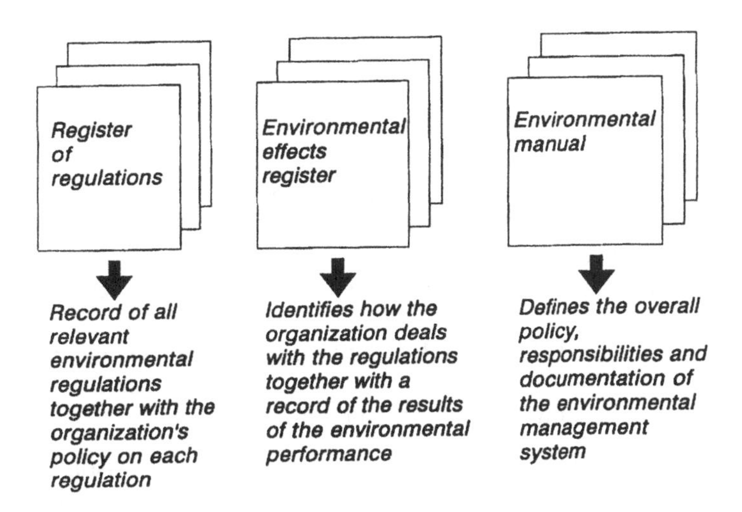

Figure 2.5 The main elements of a BS 7750 environmental management system.

reference of environmental requirements, and finally the **environmental effects register** describes the organization's response to the regulations and the records of environmental performance which are maintained.

The certification to BS 7750 is provided in the UK by accreditation agencies approved by the Department of Trade and Industry (rather than the Department of the Environment). Currently the infrastructure for the control of certification throughout Europe is still emerging and bodies currently carrying out accreditation are anticipating retrospective approval.

2.1.6. SUMMARY

- Quality management requires a 'systems' approach involving all aspects of the business in the conformance to specified requirements.
- Quality system standards such as ISO 9000 provide an excellent framework for the design and assessment of systems.
- The international developments in quality system standards throughout the 1980s represented a significant harmonization in the way quality is managed throughout the world.
- The benefits of gaining accreditation to ISO 9000 include not only the opening of new markets and the reduction in the need for multiple customer assessments but also provides the foundation for future quality development.

2.2 The scope and interpretation of ISO 9000

2.2.1 THE SCOPE OF ISO 9000

The quality system standard ISO 9000 specifies three basic levels of system.

- **Level 1** (ISO 9001, EN 29001, BS 5750: Part 1) applies where the contract specifically requires design effort and the product or service requirements are principally stated in performance terms.
- **Level 2** (ISO 9002, EN 29002, BS 5750: Part 2) applies where the specified requirements for the product or service are stated in terms of an established design or specification (which may be generated by the customer).
- **Level 3** (ISO 9003, EN 29003, BS 5750: Part 3) applies where conformance to specified requirements can be adequately established by final inspection and testing of the finished product or service.

The ISO 9000 series can therefore be seen as an 'encompassing' standard whereby all of the requirements of Level 3 are included in Level 2 and all the requirements of Levels 2 and 3 are included in Level 1:

Level 1	*Level 2*	*Level 3*
Design	—	—
Manufacture	Manufacture	—
Test	Test	Test

Sections in each part of the Standard are virtually the same so that the main differences between the levels therefore relate to the presence or absence of certain sections of the Standard. Table 2.3 provides a comparison of the parts of the Standard to illustrate the difference in scope.

The scope of the Standard, in terms of which of the elements are present and to what extent, determines how the Standard should be applied to any particular company. Typically a quality system designed to meet the requirements of Part 1 of the Standard will be in excess of 30% larger than a Part 2 application despite there being only one (5%) additional element.

Table 2.3 Cross reference of quality system elements

Clause of ISO 9004	Description of quality system element	ISO 9001	ISO 9002	ISO 9003
4	Management Responsibility	4.1	4.1*	4.1*
5	Quality System	4.2	4.2	4.2*
7	Contract Review	4.3	4.3	4.3
8	Design Control	4.4	N/A	N/A
17	Document Control	4.5	4.5	4.5
9	Purchasing	4.6	4.6	N/A
—	Purchaser Supplied Product	4.7	4.7	4.7
11.2	Product Identification and Traceability	4.8	4.8	4.8*
10	Process Control	4.9	4.9	N/A
12	Inspection and Testing	4.10	4.10	4.10*
13	Inspection, Measuring and Test Equipment	4.11	4.11	4.11
11.7	Inspection and Test Status	4.12	4.12	4.12
14	Control of Non-Conforming Product	4.13	4.13	4.13*
15	Corrective Action	4.14	4.14	4.14*
16	Handling, Storage, Packaging and Delivery	4.15	4.15	4.15
17.3	Quality Records	4.16	4.16	4.16*
5.4	Internal Quality Audits	4.17	4.17	4.17*
18	Training	4.18	4.18*	4.18*
16.2	Servicing	4.19	4.19	N/A
20	Statistical Techniques	4.20	4.20	4.20*

Key: * Requirements less stringent than ISO 9001 or ISO 9002.
N/A Element not present.

Note: The 1994 revision of ISO 9000 adopts a common numbering system for clauses in each of the three parts (e.g. in Part 2, Design is referenced as 4.4 and described as not applicable) and in addition Part 2 was extended to 19 sections to include Servicing.

2.2.2 THE APPLICATION OF ISO 9000

In adopting and applying the requirements of ISO 9000 either to the design or evaluation of a company's quality system, the following considerations need to be made.

- What level of activity applies – 1, 2 or 3?
- What structure of system should be adopted – functional or project based?

A survey of the application of ISO 9000 in the UK shows that Level 2 of the standard predominates and Level 3 is rarely used, as illustrated in Figure 2.6. The reason for this distribution of companies is due to the fact that most organizations do not design or develop their products/services on a contract-by-contract basis (hence Level 2), and other more applicable standards (for example, National Measurement Accreditation Service (NAMAS)) can be applied to companies involved primarily in testing.

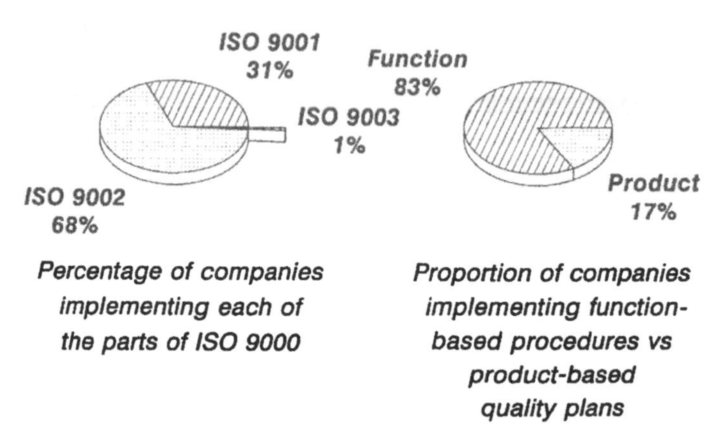

Figure 2.6 The application of ISO 9000.

The same survey also showed that most companies structure their quality systems on a functional basis rather than on a product-by-product basis, also illustrated in Figure 2.6. A functional approach means that procedures written for, say, purchasing or inspection apply to all product/services whereby a product-based structure develops quality plans in which procedures are written for particular projects

In deciding what level of activity to apply, the key distinction between Levels 1 and 2 is to what extent design plays a key role on a contract-by-contract basis. Clearly all products or services are designed at some stage but the question here is to what extent they are designed for each individual contract. Bearings or an electric kettle are clearly 'designed' products but they are, however, mostly bought against the published (for example, catalogue) specification. Most companies in evaluating which part of the Standard to apply have to consider:

- what proportion of their business requires design effort and whether they will gain business advantage from being Part 1 accredited;
- whether they should exclude certain products or sections of their business from the scope of their quality systems approval;
- in organizational terms, where they should 'bound' their system.

The following examples illustrate the approaches that should be adopted

Electrical repair company rewinding electric motors

Here the service element may indicate Part 1 of the Standard (4.19); however, this element relates to the (often on-site) servicing and maintenance of product which has been supplied. Part 3 of the Standard is not applicable because, although the company does not actually manufacture the motors, the process as described in 4.9 (Part 2) would in this case be the repair process. In addition Part 3 assessment would not control the purchasing of any replacement parts for the motor. Hence the most appropriate level would be ISO 9002.

Motor parts manufacturer

Here the company is supplying parts to a major motor car manufacturer, and although there may be a major design element in the parts to be manufactured, this design work is almost certainly undertaken by the customer who would therefore generate the parts specifications that the parts manufacturer would work to. Hence the most appropriate level would be ISO 9002.

Bulk chemical producer

Here the product design normally takes the form of an agreed international specification for standard chemicals and therefore the composition and purity are pre-defined. Hence again the most appropriate level would be ISO 9002.

International electrical/electronics products group

Here the main consideration is where to draw the boundary around the company, which may have many different subsidiaries, for the purposes of accreditation. In this situation one might assume that the group will need to have some form of all-embracing quality system covering every single group company and almost certainly to the most comprehensive Part of the Standard. The reality, however, depends upon the extent of inter-group trading as autonomous divisions of the company can be separately accredited. Hence the company may have a number of accredited group companies to ISO 9001, 9002 and 9003.

Training organization

Here the company is providing a service rather than a product and is working primarily from the basis of identifying training needs and tailoring individual or group training programmes as opposed to presenting standard training packages. Here the most appropriate level would be ISO 9001.

2.2.3 THE ELEMENTS OF ISO 9000

The Standard is an extremely concise document which has been very carefully worded and therefore needs to be read in detail to be understood. The guides to the Standard listed below are excellent documents in assisting the understanding and interpretation of ISO 9000:

- BS ISO 9000–1 *Guidelines to ISO 9000*;
- BS ISO 9000–2 *Generic Guidelines*;
- BS ISO 9000–3 *Software Guidelines*;
- BS ISO 9004–1 *System Elements*;
- BS ISO 9004–2 *Service Applications*;
- BS ISO 9004–3 *Processed Material*;
- BS ISO 9004–4 *Quality Improvements*;
- BS ISO 9004–5 *Quality Plans*;
- BS ISO 9004–6 *Project Management*;
- BS ISO 9004–7 *Configuration Management*.

A definition of the terms used in the Standard can be found in either BS 4778 or the international equivalent ISO 8402. In particular the Standard uses the following terms to describe the three components of the business chain:

- **purchaser** describes the customers of the company to be assessed to the Standard;
- **supplier** describes the company to be assessed and this reflects that ISO 9000 is primarily a supply standard;
- **subcontractor** describes the supplier of goods and services to the company to be assessed.

The other important term used in ISO 9000 is **product** which is used to describe the product or service at all stages of the process. With this interpretation raw materials, components, sub-assemblies as well as finished product are covered by the term 'product' in manufacturing, and similarly a service at all stages of provision is referred to by the generic term 'product'.

Each part of the Standard is basically divided into four sections as follows.

- **Section 1** defines the scope of application for the relevant levels of activity.
- **Section 2** details the documents referenced in the Standard.
- **Section 3** defines the terms used in the Standard by basically referencing BS 4778/ISO 8442.
- **Section 4** is a description of the quality system requirements and is divided into 20 elements in Part 1, into 19 elements in Part 2 and into 16 elements in Part 3.

A practical interpretation of the requirements of each element of the Standard together with a summary of the 1994 revision is given below.

4.1 Management Responsibility

This element requires the company to have a documented quality policy and to ensure that this is understood by all employees. Often this means issuing each employee with a copy of the policy statement or alternatively prominently displaying the policy. Organizational responsibilities need to be defined and the Standard identifies five specific areas to do with the actioning and recording of quality problems and the appropriate controls and corrective actions. Adequate resources for verification activities are required by section 4.1.2.2 and particular note should be taken of the need for trained personnel in the areas of management, performance of work and verification activities including audits. A management representative who has ultimate responsibility for the quality system needs to be identified. This is usually the quality director in the case of large companies, the quality manager in the case of medium sized companies and the general manager for small companies. The 1994 revision also requires executive responsibility to be defined together with customer expectations and needs. Managers are required to review the quality system performance periodically and this usually involves inputs from internal audits, reject records and customer complaints. The reviews normally take place approximately every six months.

4.2 Quality System

This element basically requires that the quality system should be documented in terms of policies, procedures and instructions. Guidelines are given as to the format of the documentation and the need for quality planning. The 1994 revision explicitly requires the documentation of a quality manual.

4.3 Contract Review

This element states that the customer's requirements (the 'contract') should be formally reviewed to ensure that the supplier is capable in both technical and organizational terms and that a record should be maintained of such reviews.

4.4 Design Control

This element requires that the design process should be documented and correctly planned and both the design inputs (specifications) and the design outputs (drawings etc.) should be formally documented. The supplier shall review the designs to ensure that the outputs meet the input requirements and that any changes to the designs are properly controlled. The 1994 revision requires the design reviews to be formally documented.

4.5 Document Control

The documents which form part of the quality system (procedures, specifications, detailed instructions, etc.) should be issued in a controlled way which indicates who has approved the document, the date of approval and the current level of issue revision. This ensures that everyone in the organization is working to the most up-to-date information and such documentation should be readily available to those who need to use it. Any changes to documents should also be controlled and approved by the originator and a list should be maintained of such modifications.

4.6 Purchasing

This element basically requires that the supplier should use 'assessed' subcontractors, and whilst the Standard does not compel a supplier to use only ISO 9000 approved sources (as clearly initially no company could get approval) this is an obvious and convenient approach. All purchasing documents should specify in precise terms what is required and who has approved the purchase and, where specified as part of a contract, the supplier should allow the customer to verify incoming purchases without, however, absolving the suppliers' responsibilities for checking.

4.7 Purchaser Supplied Product

Where a customer 'free-issues' equipment or materials to the supplier, the Standard requires that these materials should be properly protected and their identity maintained.

4.8 Product Identification and Traceability

This element lays down the requirement for the product or service to be correctly identified throughout the production process and that if it is required explicitly by the customer (in other words this is not mandatory) then the traceability back to constituent raw materials of batches should be maintained.

4.9 Process Control

The supplier is required to have formally documented work processes for all operations affecting quality. Process equipment and personnel should all be capable of meeting the requirements. 'Special processes' are those which cannot be examined by subsequent testing prior to use – for example, the packing of a parachute! Such processes should have particular attention paid to their capability and control and to the standards of workmanship required. For the first time, the 1994 revision also requires the provision of equipment maintenance.

4.10 Inspection and Testing

The supplier should ensure that incoming 'goods' are verified (not specifically tested, so the Standard does not compel 100% inspection) before use and, if required urgently, then a concession is required so that goods are capable of being recalled. Furthermore, the product or service should be tested at all stages of the process laid down in the quality plan and final release should only be allowed when the finished product has met the specified test requirements. Records shall be kept of the testing undertaken indicating the person(s) responsible for testing and the acceptance criteria applied.

4.11 Inspection, Measuring and Test Equipment

Affectionately known as the 'calibration' section, this element of the Standard basically requires that the equipment used to test the conformance of the product is in good working order and in a known state of calibration. The Standard requires that measurement certainty be known, which is often difficult with certain subjective tests, and that the calibration should be traceable to established national or international standards. The Standard does not specify which equipment should be calibrated nor how often. The periodicity of calibration is regulated, however, by the requirement to assess previous inspection data when equipment is found to be out of calibration. Equipment should be properly identified and protected and records of the calibration information should be maintained.

4.12 Inspection and Test Status

This element requires that the supplier shall identify whether a product has been tested or not and if it has been tested whether it has passed or failed. Many different approaches to identifying test status are allowed including labels, segregation, test records, etc.

4.13 Control of Non-Conforming Product

Product which is out of specification shall be controlled and not used and the supplier shall review the product with a view to either rejecting/reworking, providing a concession or regrading. The customer can require in the contract to be told of any concessions and any rework shall be undertaken in line with documented procedures.

4.14 Corrective Action

This element of the Standard requires the supplier to continuously review and amend the quality system in the light of non-conformances. The supplier shall review what went wrong, why and what to do to prevent recurrence and must also analyse the system regularly to detect and correct potential non-compliances.

4.15 Handling, Storage, Packaging and Delivery

Sometimes referred to as the 'housekeeping' section, this part of the Standard requires that the product shall be handled and stored in such a way as to prevent damage or deterioration. The product shall be packaged in conformance with requirements and, depending upon the contractual arrangements, the supplier shall also be responsible for protection of the product during delivery.

4.16 Quality Records

The records of the performance of the quality system (contract reviews, subcontractor assessments, test records, calibration data, etc.) shall be maintained in a readily accessible format for a prescribed period of time. The Standard does not, however, specify how long records should be kept as clearly the retention period for the test results for, say, biscuits with a shelf life of a matter of weeks would be very different from the commissioning data from a nuclear power station with a working life in excess of 20 years.

4.17 Internal Quality Audits

The Standard requires the supplier to 'police' their own quality system to ensure the laid-down policies, procedures and instructions are being followed. These audits should be undertaken by suitably qualified employees who are independent from the area being audited. Audits shall be carried out at appropriate frequencies (typically three to six months) and the results reported to the managers responsible for the area audited to take any corrective actions identified.

4.18 Training

Employees whose work affects quality need to be properly trained. The phrase 'personnel affecting quality' is interpreted quite liberally and the Standard states that the training can either be in the form of educational background, training courses attended or experienced gained. The supplier should be able to demonstrate through documented records the training of such staff and must show that the training requirements are reviewed regularly.

4.19 Servicing

Where a contract requires the product to be maintained after it has been supplied then again the supplier should have a documented method describing such servicing. These procedures will apply even if the servicing is carried out by a third party, for example in the case of an automotive garage. The requirements for servicing were introduced in Part 2 of the Standard (ISO 9002) in the 1994 revision

4.20 Statistical Techniques

Where the supplier uses data analysis methods, for example in market analysis, process control, reliability analysis, product sampling, etc., then the methods used should have a valid statistical basis and again shall be carried out in accordance with documented procedures.

2.2.4 SUMMARY

- ISO 9000 specifies three levels of system and identifies up to 20 elements which define quality assurance management.
- In selecting the most appropriate level of system a company must decide to what extent contracts require design effort. Most companies are assessed under ISO 9002.
- The individual elements of the Standard define the requirements of a supplier's quality system covering everything from responsibilities, controls and records to workmanship.

2.3 The implementation of ISO 9000

2.3.1 THE STAGES IN THE IMPLEMENTATION OF ISO 9000

Most of the companies which have been accredited to ISO 9000 have undergone some form of implementation process whereby new quality system elements have been introduced within their organizations. All organizations employ a quality system to some extent but in order to achieve accreditation to ISO 9000 the system generally needs to be structured and presented in line with the requirements of the Standard.

The basic stages in the implementation of ISO 9000 are:

1. to define the quality objectives of the company and establish the existing systems and procedures;
2. to develop the existing system and introduce new elements in line with both the business objectives and the requirements of ISO 9000.

The mistake many companies make in the implementation of ISO 9000 is to attempt to adopt some form of 'off-the-shelf' documented quality system rather than to develop and improve their existing business processes.

In terms of the detailed activities involved in developing quality systems to ISO 9000 these may be summarized as follows.

- **Analysis** involves an identification of the quality objectives, a review of the existing quality systems, the securing of the commitment of senior management and the development of the ISO 9000 implementation plan.
- **Product or service specification** involves developing contract review procedures, design and development procedures, the procedure for controlling the issue, approval and change of the quality system documentation to be generated, and the production of quality plans.
- **Material control** requires procedures for the specification of bought-in goods or services, methods for assessing subcontractors and procedures for receiving raw materials, including sampling plans and control procedures for any materials supplied directly by the customer.
- **Process control** requires procedures for identifying the product throughout the conversion process and for maintaining traceability. Procedures are also required to describe the control of all the main production or business processes and any associated detailed work instructions. Methods should also be developed to control any non-conforming material throughout the process and standards of workmanship need to be defined together with criteria for acceptability of the product or service at the appropriate process stage.
- **Inspection and testing** includes the methods for in-process and final checking of the product or service contract including the definition of acceptance criteria. Details are required of how the inspection status of

the product is maintained and procedures for the calibration of equipment and the associated records. This stage should also include details of any statistical techniques used.

- **Quality records** require procedures for the maintenance and storage of quality system records and also procedures for the periodic auditing of the system. Procedures are also required for the recording of employee training.
- **Quality manual** should be a relatively brief document (20 or so pages) stating the business policy with respect to quality and should be structured in line with the sections of ISO 9000. Detailed procedures should be cross-referenced in the quality manual which should be seen as a means of promoting the company's approach to quality.

The advantage of this approach whereby the implementation is broken down into elemental stages is that the tasks can then be delegated to the appropriate managers responsible which promotes both the accuracy of the procedures and the organizational ownership.

The timescales for implementing ISO 9000 vary considerably depending upon company size and system complexity and also the organizational resources and commitment provided. The timescales identified by the research described above are shown in Figure 2.7.

Similarly the cost of implementing ISO 9000 will vary from company to company and typical costs are shown in Figure 2.8. The major costs involved include management time in the preparation of documentation, the provision of secure storage areas, calibration costs including test equip-

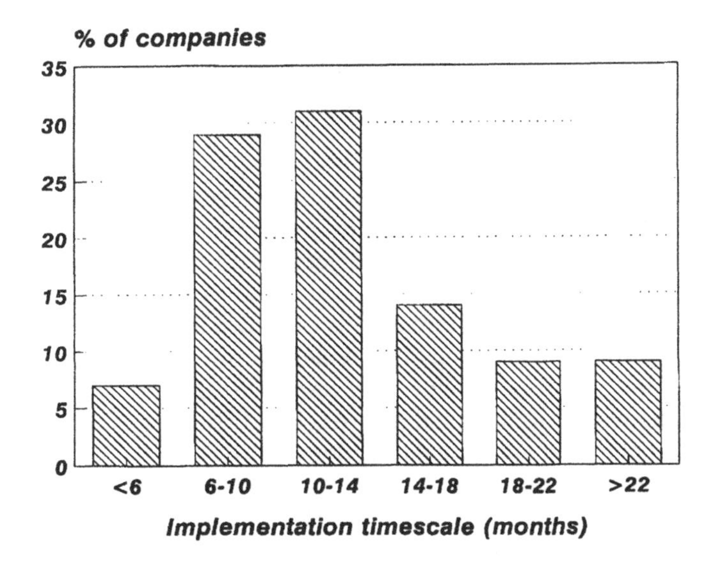

Figure 2.7 Typical ISO 9000 implementation timescales.

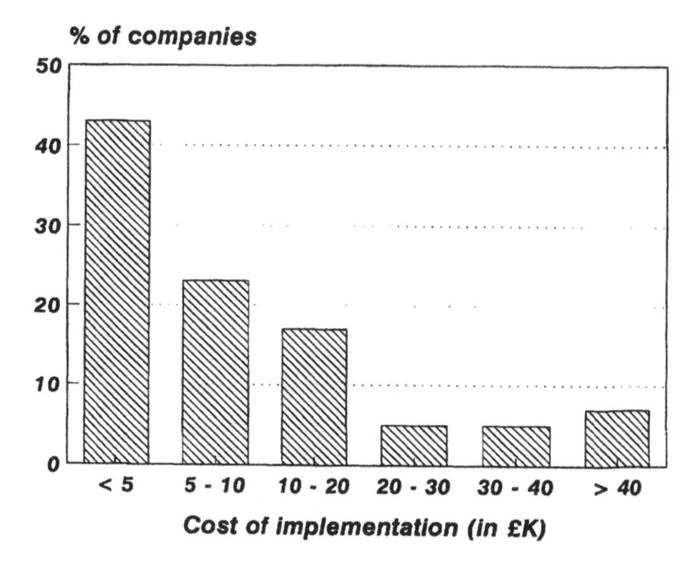

Figure 2.8 Typical ISO 9000 implementation costs.

ment, and assessment costs. Some companies, although by no means all, also require additional personnel to either manage or operate the quality system.

2.3.2 QUALITY SYSTEM DESIGN AND STRUCTURE

The formalized design of the quality system is a process which most companies undertake only once (ignoring subsequent modifications and amendments) and for this reason the process often benefits from the advice of an external consultant. In the UK this has been recognized under the Department of Trade and Industry's Enterprise Initiatives which have provided consultancy support for small/medium companies.

There exists as many different quality systems as there are companies, each one reflecting both the business objectives of the individual company and the requirements of ISO 9000. A basic design framework to adopt is to structure the quality system at three levels as shown in Figure 2.9.

The first level of documentation, widely termed the quality manual, should state the company's policy with respect to Quality and identify the corresponding responsibilities and records. This first level of documentation should be structured in line with the sections of ISO 9000.

The third level of documentation is normally company-specific giving details of how individual work processes (for example, welding, soldering, testing) are carried out within the company. These work instructions should in addition to specifying **how** the work should be done should also identify

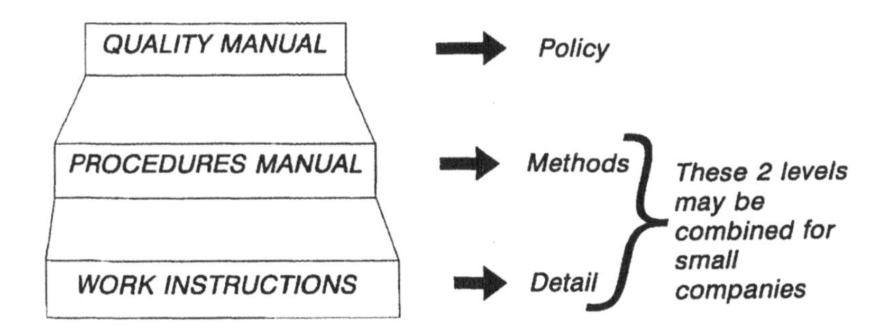

Figure 2.9 Basic quality system structure.

who should undertake the work and **what** records are maintained. Work instructions should be clear, concise and auditable.

At the second level of documentation, quality procedures, opinion is divided as to whether they should reflect the requirements of the Standard or the operation of the business. Most companies employ a hybrid approach whereby certain procedures relate specifically to the Standard (for example, 'Control of Non-Conforming Product') whereas others reflect the business processes (for example, 'Preparation of Quotations').

Certain elements of the ISO 9000 quality system framework lend themselves to generic designs. For example:

- **Vendor assessment** requires some form of supplier identification/categorization, a definition of the supply specification together with a record of the supplier assessment, as shown in Figure 2.10.
- **The calibration system** generally requires some form of gauge identification, a schedule for the calibration of equipment and finally individual calibration records showing details of the calibration method, accuracy, frequency etc., as shown in Figure 2.11.
- **The training** element should contain details of the overall training requirements of the quality system to identify which personnel have been trained and also the individual training records showing information on the training undertaken, as illustrated in Figure 2.12.

2.3.3 PREPARATION OF QUALITY SYSTEM DOCUMENTATION

The procedures and work instructions written to describe the quality system should be clear, concise and auditable. In preparing the documentation, the first task is to establish the format for the documents to facilitate control purposes. This normally requires some form of document reference and issue status together with a document authorization mechanism as shown in Figure 2.13. Some form of master list of procedures is normally maintained

XYZ Engineering Ltd High Street Newtown		Ref: SA/02 Issue: 01 Date: Jun 94
SUPPLIER ASSESSMENT RECORD Supplier category: A Supplier name: ABC Chemicals, #1 Ind. Est., Seaport Ref: ABC 1 Material: Sulphuric acid (Spec. Ref. XYZ 07)		
Non-conformance	Date	Corrective action

Figure 2.10 Vendor assessment system.

XYZ Engineering Ltd High Street Newtown			Ref: EM01 Issue: 01 Date: Jun 94
CALIBRATION RECORD Equipment: External micrometer 0–25 mm Manufacturer: Mitutoya Calibration method: Internal calibration using slip gauges (Ref SG01) at 0, 5, 10, 15, 20, 25 mm; external using Measuring Equipment Services Ltd Required accuracy: ± 0.0025 mm Calibration frequency: Internal (2 months); external (12 months)			
INTERNAL	EXTERNAL		Accuracy/comments
Date	Date sent	Date recd.	

Figure 2.11 Calibration system record.

XYZ Engineering Ltd High Street, Newtown					Ref: TR/02 Issue: 01 Date: Jun 94
TRAINING RECORD Employee name: G. Smith Employee ref: ABC/123					
Training	Date review	Date start	Date finish	Comments	Signature
Yamazaki machining centre operations	Mar 95	Oct 93	Mar 94	Complete	PC
Inspection operations	Dec 95	Jun 93	Sep 93	All gauges	PC
Quality systems training	May 95	May 93	2 day	Quality college	PC

Figure 2.12 Quality system training record.

XYZ ENGINEERING CO. LTD PROCEDURE TITLE: Procedure for the control of	PROCEDURE REF: QAP 001 ISSUE: 01 DATE: 01/06/94 PAGE: 1 OF 3
PROCEDURE DETAILS	
PREPARED BY:	APPROVED BY:

Figure 2.13 Typical format for quality system documents.

which records the procedures in use, the current revision status of each procedure and the circulation lists.

The basic structure of the procedures should be as follows:

- **purpose** which describes the activity covered in the procedure;
- **scope** which describes which aspects of the activity are covered in the procedure;
- **related documents** which are those documents referred to by the procedure;
- **procedure** which describes the detail of the element;
- **documentation** which describes the documents produced in the execution of the procedure and the records required.

Consideration as to how the procedure is to be subsequently audited should be a key concern in developing the documentation and in particular attention should be given to:

- responsibilities as to who should carry out the procedure;
- description of any specifications, settings or acceptance criteria;
- details of the records which should be taken to illustrate conformance to specification;
- any special skills assumed or training required.

When planning the quality system documentation, attention should be given to the definition of which procedures are required (an overall system game plan) in order to eliminate potential overlap or, alternatively, a ballooning effect whereby procedure after procedure is added to what appears to be an incomplete 'masterpiece'. The best approach is to encourage those (managers) responsible for the process to draft the procedure as this ensures:

- detailed understanding of the process described;
- ownership of the completed procedure.

In addition it is often good policy to circulate the procedure for comment to the employees responsible for carrying out the procedure.

In developing a quality system procedure it is normally good practice to create a flowchart of the process to be described. A common mistake is for authors to over-elaborate and to write procedures in what they perceive to be a required 'quality assurance speak'. This produces unworkable controls and unnecessary bureaucracy – a criticism often levelled at ISO 9000 but in almost all cases self-generated.

2.3.4 EXAMPLES OF ISO 9000 QUALITY SYSTEM IMPLEMENTATIONS

Many different structures and formats of quality system have been accredited to ISO 9000 and therefore there is no single, successful approach.

Individual procedures, however, should follow the format described in section 2.3.3, an example of which is shown in Figure 2.14. The procedures are usually either:

- quality-specific procedures, for example Calibration, or
- business procedures, for example Purchase Order Processing.

The document describing the quality policy, usually termed the quality manual, should be seen as an overview document which can be used to promote the company's quality system externally. The quality manual generally

XYZ Engineering Ltd | *Proc: QAP #*
Issue: XX
Procedure: Receiving Inspection | *Date: YYYY*

RECEIVING INSPECTION

1. Introduction
   1.1 This document details the operation of (inwards goods inspection) in the inwards goods receiving area.
2. Scope
   2.1 This procedure applies to all (inwards goods inspection) activities in the .... site.
3. Related Documents
   3.1 Company Documents
   (List of relevant company procedures/standards).
   3.2 Customer/National/International Documents
   (List of relevant documents).
4. Receiving Documentation
   4.1 On receipt a (Goods Received Note) will be raised – form (ABC 123) refers.
5. Inspection Routines
   5.1 On receipt of the (GRN) the (inspection area) will examine the goods or materials against the purchase order and any relevant specifications.
   5.2 If the goods are acceptable they will be passed on to the next stage or process and the (GRN) marked accordingly.
   5.3 If the goods are defective the (GRN) will be marked accordingly and the goods physically identified and segregated. The procedure for handling defective goods is contained in QAP #.
6. Records

PREPARED BY: APPROVED BY:

Figure 2.14 Sample quality system procedure.

does not contain a great deal of proprietary detail describing how key processes are carried out but instead will refer to specific quality procedures or work instructions.

As described in section 2.3.2, the quality manual should be structured in line with the elements of ISO 9000 as this assists the assessors in the comparison with the requirements of the Standard. A typical quality manual layout is shown in Figure 2.15.

COMPANY QUALITY MANUAL ISO 9002 – 1994

0. INTRODUCTION
1. SCOPE AND FIELD OF APPLICATION
2. REFERENCES
3. DEFINITIONS

   QUALITY POLICY STATEMENT

4. QUALITY SYSTEM REQUIREMENTS
   - 4.1 Management responsibility
   - 4.2 Quality system
   - 4.3 Contract review
   - 4.5 Document control
   - 4.6 Purchasing
   - 4.7 Purchaser supplier product
   - 4.8 Product identification and traceability
   - 4.9 Process control
   - 4.10 Inspection and testing
   - 4.11 Inspection, measuring and test equipment
   - 4.12 Inspection and test status
   - 4.13 Control of non-conforming product
   - 4.14 Corrective and preventative action
   - 4.15 Handling, storage, packaging & delivery
   - 4.16 Quality records
   - 4.17 Internal quality audits
   - 4.18 Training
   - 4.19 Servicing
   - 4.20 Statistical techniques

APPENDIX – List of quality system procedures

Figure 2.15 Structure for a quality manual.

In terms of the overall structure of an individual company's quality systems a number of different approaches have been adopted whereby the business detail is described in different ways and at various levels. Typical examples of quality system layouts are shown in Figure 2.16.

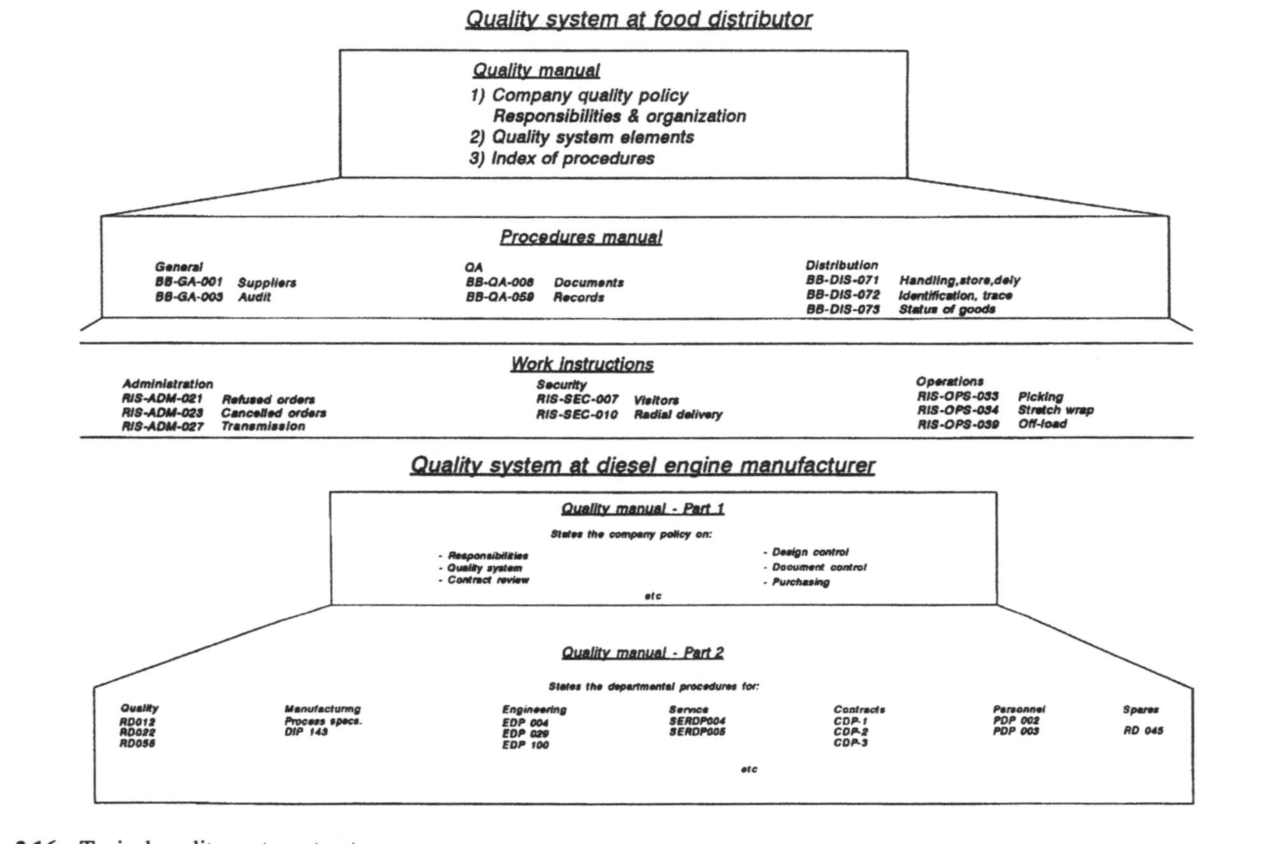

Figure 2.16 Typical quality system structures.

2.3.5 SUMMARY

- Introducing ISO 9000 quality management systems requires the documentation and implementation of quality policy, procedures and practices and typically takes 12 to 18 months.
- Many different designs of quality systems have been accredited to ISO 9000 although generally three levels of documentation are employed – policy, procedures and work instructions.
- Quality system documents should have a format which facilitates issue control and should be created by the owners of the process described.
- Quality manuals should be structured in line with the elements of ISO 9000 and procedures describing the operation of the quality system should be clear, concise and auditable.

2.4 Quality system audit and maintenance

2.4.1 THE STAGES AND DEPTHS OF QUALITY SYSTEM AUDITS

An audit is a systematic and independent examination to determine whether quality activities comply with planned arrangements. The reasons for auditing quality systems include:

- internal audits are an explicit requirement of ISO 9000 (4.17);
- audits are required to provide objective evidence that the quality system is functioning and assist in the 'policing' of the system to ensure the system does not deteriorate;
- internal audits provide important input to the management review of the performance of the quality system in achieving the quality objectives.

The main stages of an audit are as follows:

- planning – 40%
- performing – 40%
- reporting – 10%
- follow-up – 10%

The percentages indicate the approximate amount of time to be devoted to each stage.

Audits can be carried out at different depths of detail depending upon the objectives of the audit. Basically the audit can be carried out at a 'systems' level which involves a broad evaluation of all elements of the quality system described in the quality manual or at a 'compliance' level which provides an in-depth evaluation of specific quality activities and procedures, as shown in Figure 2.17.

In terms of the organization carrying out the audit, the following distinctions can also be made.

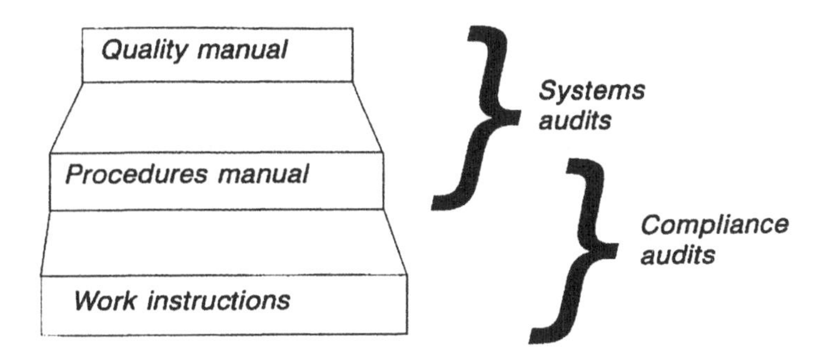

Figure 2.17 Depth of quality system audits.

- **internal audits** are carried out by a company on its own quality system.
- **External audits** are carried out by a company to evaluate suppliers' quality systems.
- **Extrinsic or second/third party** audits are carried out on the company by a customer or third-party organization.

Based upon the above definitions the most common forms of audits therefore are:

- internal company audits – internal, compliance audits;
- assessment of suppliers – external, systems audits;
- accreditation assessments – extrinsic, systems audits.

2.4.2 AUDIT TOOLS AND TECHNIQUES

The two key viewpoints necessary when carrying out an audit are:

- a clear understanding of the detail of the procedures being audited;
- a clearing understanding of the requirements of ISO 9000.

In terms of each of the main stages of the audit the key tools and techniques are as follows.

Audit planning

The key elements here are as follows:

- **Audit schedule** – a matrix of the timing as to when each audit element is to be checked throughout the year.
- **Audit personnel** – the appointment of an auditor or, where the scope of the audit is considerable, an audit team leader and team members.
- **Notification of the auditee** – request for all the quality system documentation relevant to the audit. Many people feel you should surprise an auditee – this is universally the wrong approach.

- **Preparation of checklists** – lists of specific audit questions which are used during the audit to prompt the auditor. Through the use of checklists the auditor can ensure the pace and completeness of the audit and also record the audit findings. Well prepared checklists are the key to effective quality system audits.

Audit performance

The key techniques here are as follows.

- **Entry meetings** to initially brief the auditee and arrange escorts, and exit meetings to finally provide feedback on the audit findings.
- **The audit** itself is a very practical (and often unpredictable!) experience. Auditing is normally conducted on a sampling basis (illustrative records are examined) and objective evidence should be sought and recorded (both the affirmative and deficient findings). Checklists should be used as a guide and expanded if necessary
- In terms of asking **audit questions**, Rudyard Kipling's rhyme should be borne in mind: 'I keep six honest serving men, they taught me all I knew – their names were *what* and *why* and *where* and *when* and *how* and *who*.' In other words open questions should be asked.
- **Any deficiencies** noted during the audit should be confirmed with the auditee, and not come as a surprise later.

Audit reporting

This primarily involves the recording of any non-conformances and summarizes the audit findings for purposes of management review. The key elements are:

- **corrective action requests (CARs)** which detail concisely any discrepancies found and require the auditee to complete details of the proposed corrective actions, together with timescales;
- **audit summaries** which should detail the general findings of the audit together with the required follow-up action.

Audit follow-up

The key technique here is:

- **the audit close-out** which ensures any CARs raised during the audit have been corrected and a record of this made.

If CARs are not closed out then the action should be escalated which normally involves referring to a higher level of management.

2.4.3 AUDIT DOCUMENTATION

An explicit requirement of ISO 9000 is that a record should be kept of the internal audits. The key records which are normally expected are those associated with the audit planning (schedules and checklists), details of the corrective actions identified and records of the reporting of the audit to the quality system manager.

The key documents are as follows:

- **The audit schedule.** This details the timetabling of the audit as shown in Figure 2.18. Elements should be scheduled according to their importance to the effective functioning of the quality system. Audit schedules can, however, be amended in the light of audit findings and the frequency of auditing certain elements can be either increased or decreased.
- **Audit checklists.** These provide a document for recording the questions to be asked during the audit and are prepared by the auditor reading through the detailed procedure. The format for an audit checklist is shown in Figure 2.19. The checklist should be used as a guide to the auditor during the conduct of the audit and a record should be made as to whether the element in question, for example 'Are the correct test records being maintained?', is being complied with. Any additional comments or observations should also be recorded. The checklists should not, however, limit the scope of the audit and cannot be used as a substitute for observant, focused and detailed auditing skills.
- **Corrective action requests.** These record any non-compliances found during the audit and give details of the corresponding corrective actions, as shown in Figure 2.20.

  In addition to describing the proposed corrective action, the auditee should also provide details as to how the non-compliance is to be prevented from recurring together with the date of completion for the corrective action. The auditor should then revisit the area after the completion date in order to verify the corrective action and close out the audit.
- **The audit report.** This provides a summary of the audit findings as illustrated in Figure 2.21. The audit report should also record the positive findings and areas in which the quality system is operating effectively. The audit report is normally the primary record of the audit.
- **Management review reports.** These provide details of the relative performance of various elements of the quality system, as shown in Figure 2.22. The Management reports normally summarize the non-compliances both in terms of the quality system element but also in terms of the department or business function responsible for the discrepancy.

System Element	Proc. Ref.	J	F	M	A	M	J	J	A	S	O	N	D
4.1	QAP 1	X						X					
4.2	QAP 2	X						X					
4.3	QAP 3		X						X				
4.4	QAP 4		X						X				
4.5	QAP 5			X						X			
4.6	QAP 6			X						X			
4.7	QAP 7			X						X			
4.8	QAP 8			X						X			
4.9	QAP 9-12				X						X		
4.10	QAP 13 -16				X						X		
4.11	QAP 17					X						X	
4.12	QAP 18					X						X	
4.13	QAP 19					X						X	
4.14	QAP 20					X						X	
4.15	QAP 21						X						X
4.16	QAP 22		X						X				X
4.17	QAP 23						X						X
4.18	QAP 24						X						X
4.19	QAP 25						X						X
4.20	QAP 26						X						X

Figure 2.18 Audit schedule sheet.

Check list for audit report no. Page of Procedure ref: Auditor: Date:					
Item	Requirement	Proc. ref.	Sample size	Audit results	Comments/remarks
				A	Satisfactory
				B	Minor non-compliance
				C	Major non-compliance

Figure 2.19 Audit checklist.

Corrective action request		
CAR No:		
Company/department:		Audit no:
Audit basis/procedures:		Date:
Auditor:	Auditee:	Dept:
Non-compliance:		
Signed (auditee)	Signed (Auditor)	
Proposed corrective action		
Signed (Auditee)	Proposed date:	
Follow-up/close out:		
Close out date:	Signed (Auditor)	

Figure 2.20 Corrective action request form.

Audit report summary sheet	*Audit report no.*
Company........ *Location:*........ *Company rep:*........	*Audit scope:*........ *Audit date:*........ *Audit team:*........
Summary of audit	
Total non-compliance reports:........ *Date of follow-up audit:*........ *Signature (Auditor):*........	

Figure 2.21 Audit summary report sheet.

2.4.4 ISO 9000 ACCREDITATION

Accreditation can be undertaken by any one of the NACCB approved bodies. The basic stages to accreditation are as follows:

System implementation

The company designs and implements its quality system and carries out a complete set of internal audits. There is a normal expectation of at least three months of evidence of the quality system operation prior to the accreditation visit although this can present problems where process cycles (for example, a product design cycle) may take considerably more than three months and therefore all stages of a complete process may not be documented. This is sometimes overcome by sampling sequential stages from different (say, design) projects.

System element	SALES	DESIGN	INSPECTION	PURCHASING	ENGINEERING	PRODUCTION	TOTAL
4.1				1			1
4.2		1			2		3
4.3	4	5					9
4.4		6					6
4.5	1	2		4	4	1	12
4.6				3			3
4.7							0
4.8						2	2
4.9						4	4
4.10			5				5
4.11			2		1	2	5
4.12			1				1
4.13			2	1			3
4.14						2	2
4.15							0
4.16		1	2				3
4.17						2	2
4.18						2	2
4.19							0
4.20							0
TOTAL	5	15	12	9	7	15	63

Figure 2.22 Management review report sheet (figures indicate the number of non-compliances or deficiencies found in each area).

Document review

The quality system documentation is forwarded to the accreditation agency for review against the requirements of ISO 9000. It is normal at this stage to forward the quality manual, procedures manual and a sample of work instructions. The accreditation agency will then notify the company of any discrepancies (usually requirements of ISO 9000 not covered by the company's documented systems) which will need to be addressed prior to the accreditation visit.

Accreditation audit

This is normally carried out by audit team of qualified lead assessors who ensure that:

- the requirements of ISO 9000 are being met in practice;
- the company's quality system is being implemented as documented.

The accreditation auditors adopt a very formal approach and set the scope of the audit at the entry meeting, basically following the agenda laid down by the relevant part of ISO 9000. Any non-compliances are identified at the time of the accreditation visit and normally classified as either major (absence or total breakdown of a system element) or minor (isolated incidence of non-conformance).

Accreditation report and surveillance

The company is normally informed at the completion of the accreditation audit whether they have been successful. Companies can fail due to:

- a major non-compliance;
- large numbers of minor non-compliances.

Most companies quality systems do have a small number of minor non-compliances identified at the accreditation visit and are given (normally) 30 days to inform the accreditation agency of appropriate corrective action. Upon receiving accreditation, which is normally granted for three years, the company can advertise its quality system as being to ISO 9001, 9002 or 9003 (but cannot advertise that its products conform to ISO 9000). The accreditation agency will then carry out surveillance audits approximately every six months to verify that the quality system is being maintained and issues non-compliance reports as appropriate. The cost in terms of the damage to a company's quality image means that very few organizations have allowed their quality systems to degenerate and thereby lose their accreditation to ISO 9000. Surveillance visits normally focus upon areas of weakness identified during the assessment or previous surveillance visits and provide the basis for continual improvement to the system.

2.4.5 SUMMARY

- Quality systems are maintained through a process of auditing which requires careful planning prior to the execution which can be either systems level or compliance level.
- Key techniques in auditing are the preparation of checklists prior to the physical audit, effective questioning to obtain objective evidence and corrective action reporting.
- Audits are documented using defined schedules, checklists, corrective action requests, summary reports and management review reports.
- The cycle of accreditation to ISO 9000 involves a review of the system documentation, a physical audit, an audit report review and, if successful, surveillance audits approximately every six months.

3 Quality costs and performance measurement

3.1 The philosophy of the cost of quality

3.1.1 THE ECONOMICS OF QUALITY

Fundamental to the idea that quality development is a critical dimension of competitiveness is the notion that the costs are outweighed by the business benefits. Foremost in the minds of managers when considering implementing quality improvement are the questions:

- What is the cost of quality management?
- What are the returns from quality improvement?

Perhaps the most powerful message that emerged during the 1970s and 1980s in relation to the importance of managing quality effectively was:

> Quality development can simultaneously improve the *external* economics of a company (in terms of market share/revenue) and the *internal* economics (in terms of the cost of providing the product or service).

It is often remarked that in business there are only two problems – **customers** and **competitors** – and therefore the attractiveness to a chief executive of a process that at the same time both improves the satisfaction of customers and reduces costs in comparison with competitors is overwhelming.

If quality is viewed as the 'conformance to specifications' then in essence the economics of quality relates to the balance between:

- the cost of conformance; and
- the cost of non-conformance.

In terms of the overall economics of quality, the message is a simple one:

> *No* product or service is ever cheaper to provide through doing things wrong the first time. Failure is expensive and often unmanageable, whereas prevention can be a planned management investment with significant overall cost benefit.

However, to fully understand the 'micro' economic processes within an individual business in terms of how certain quality improvement activities translate into cost benefits requires careful analysis. Through investing in the cost of conformance, we can reduce the cost of non-conformance and therefore reduce the overall cost of ownership as shown in Figure 3.1.

It is the positive relationship that exists between the investment in quality improvement and the benefits from reduced operational costs that has prompted commentators such as Crosby to observe that 'Quality is free.'

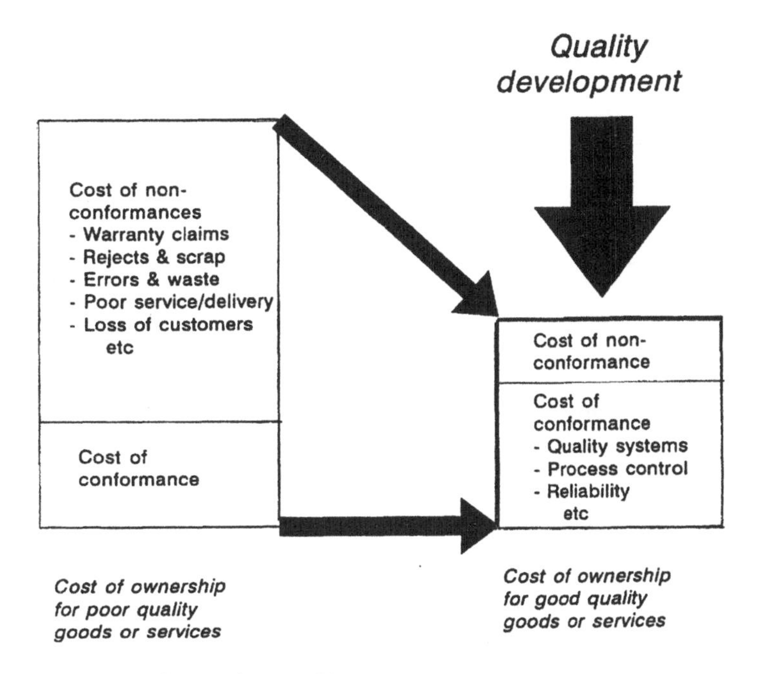

Figure 3.1 The total cost of ownership.

3.1.2 TYPICAL QUALITY COSTS

It is the magnitude of quality costs that has provided so much of the impetus to the 'quality revolution' in manufacturing and service companies. Independent surveys from many different countries have shown:

- Manufacturing companies typically can spend between 20% and 30% of total sales revenue on the cost of quality;
- service companies spend in excess of 50% of their operational costs on the management of quality.

Such statistics may be reflected upon as facts of industrial life if it were not for the fact that many of the mature quality-orientated companies are quoting quality costs at less than half these industry norms.

Taking the categories of quality costs (section 3.2.2) as:

- prevention costs;
- appraisal costs;
- failure costs;

the typical distribution of quality costs through the quality development lifecycle are as shown in Figure 3.2. Represented in this way, the economics of quality is an attractive proposition, particularly in mature markets where the potential for increasing profitability through increased market share is limited.

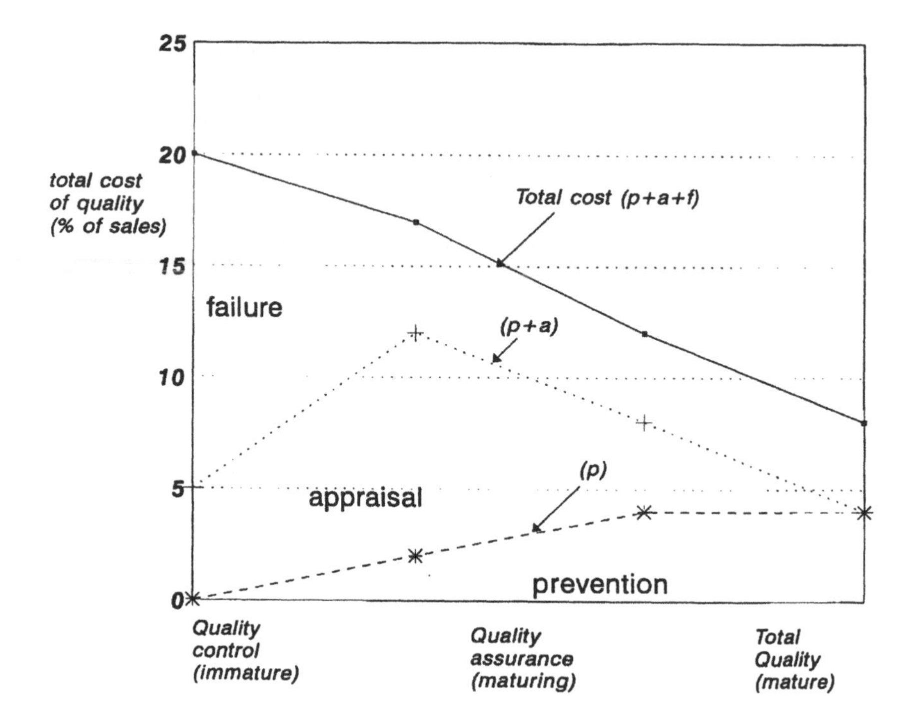

Figure 3.2 The distribution of the cost of quality with quality development.

However, two cautionary features should be borne in mind.

- The cost of quality function shown in Figure 3.2. represents an evolutionary model over a number of years of quality development, rather than an instantaneous relationship in which a company could decide at any point in time where it wished to be on the curve.

- Some of the cost benefits associated with quality development are extremely difficult to measure in traditional accounting terms. Customer loyalty, staff commitment and product or service reputation are often intangible but important business benefits of an improvement in quality performance.

Despite these problems of quantification, the concept of measuring the cost of quality is an important 'bridge' between the evangelistic approach whereby the need for quality development is accepted as an act of faith and the realities of a business world whereby quality initiatives need to demonstrate measurable improvement in profit performance. Nothing attracts the attention of senior managers to the philosophy of quality development like the promise of monetary benefit!

3.1.3 THE ROLE OF QUALITY COSTS IN THE QUALITY DEVELOPMENT PROCESS

Most organizations experience problems in introducing the concept and the measurement of quality costs, as this approach differs from traditional cost management reporting. These problems arise from the following.

- The difficulty in measuring and quantifying certain quality parameters such as the true cost of errors in the organization which are not always recorded or the improved workmanship of a correctly trained employee.
- The apparently subjective nature of the classification of quality costs and therefore the importance of establishing a consistent, company-wide view of the various quality costs.
- The technical accounting problems of a cost measurement system which by definition is incomplete (not all costs within an organization are classified as quality costs) and therefore does not produce a 'balanced' account which would permit validation.
- The difficulties in apportioning overheads across the quality cost categories. This represents a fundamental problem as the overhead activities (such as management time spent correcting errors and troubleshooting) are often the key quality costs. The moves towards activity-based systems may improve this particular difficulty.

Despite these problems, the ongoing measurement of the cost of quality is an important element of quality development. The main benefits in adopting a system for measuring the cost of quality are:

- **management focus** in providing a measure of financial benefit to sceptical senior managers struggling along the demanding road of quality improvement;
- **project focus** in helping to identify which areas of the business should be prioritized in terms of quality improvement activities;

- **measurement focus** in establishing a quantifiable performance measure of the progress made through the more effective management of quality.

The reporting format and frequency are, generally, company-specific although certain 'standardized' approaches are emerging as discussed in section 3.2. The key indices normally quoted for the cost of quality are as follows.

- **Sales ratio.** This is the ratio of the total cost of quality to the net sales value (usually expressed in financial terms and reported to senior management):

$$\frac{\text{Total quality cost}}{\text{Net sales}}$$

- **Cost ratio.** This is the ratio of the total cost of quality to the overall cost of operation:

$$\frac{\text{Total quality cost}}{\text{Cost of operation}}$$

- **Unit ratio.** This is the ratio of the cost of quality per unit produced (manufacturing) or provided (service):

$$\frac{\text{Total quality cost}}{\text{Units produced}}$$

- **Labour ratio.** This is the ratio of the cost of quality to the direct cost of labour (usually expressed in hours and reported to line management):

$$\frac{\text{Total quality cost}}{\text{direct labour costs}}$$

The most widely used of these cost quality bases is the sales ratio. This is because this ratio represents both the internal implications of quality management (reduced cost of quality) and the external implication (improved customer service). However, like all simple measures of complex activities, such ratios should be treated with care to ensure that any 'special factors' such as labour automation, product mix changes or transient market conditions do not distort the measures.

3.1.4 SUMMARY

- The economics of quality management are such that through the investment in prevention, the overall cost of quality to the business can be significantly reduced in the early stages of quality development.

- For immature organizations the cost of quality can be typically 20% of sales income.
- Quality costs can be used to focus management attention to the importance of quality management and to financially measure the benefits of quality improvement.

3.2 Methods for collecting quality costs

3.2.1 PROCESS COST MODELS

Two basic approaches to the collection and presentation of the cost of quality have emerged:

- prevention, appraisal, failure (PAF) costing;
- process cost modelling.

Both of these methods are well defined and in the UK they are published in the form of a British Standard BS 6143:

- BS 6143 Part 1: 1992 *Process Cost Model*;
- BS 6143 Part 2: 1990 *Prevention Appraisal and Failure Model.*

The more traditional of the two approaches was the PAF costing model, but increasingly as companies move towards total quality management (Chapter 5), the need to adopt a business process viewpoint has increased the attractiveness of process cost modelling.

In determining the cost of quality for a business process, the basic elements need to be defined as shown in Figure 3.3.

To identify the cost of quality using the process cost approach the key stages are:

- **Stage 1** – establish the quality improvement team to prepare the process cost model and collect the data.
- **Stage 2** – identify the business processes for which the cost of quality is to be determined.
- **Stage 3** – decompose the business process in terms of the constituent key activities and for each of these activities identify the costs associated with people, equipment, materials or environment.
- **Stage 4** – classify the quality costs in terms of the cost of conformance or the cost of non-conformance.

The costs collected can be either actual costs taken from financial reports or synthesized costs which are derived from elemental data (for example, the number of hours taken to complete an activity × hourly rate). The elements of a typical process model cost of quality report is shown in Figure 3.4.

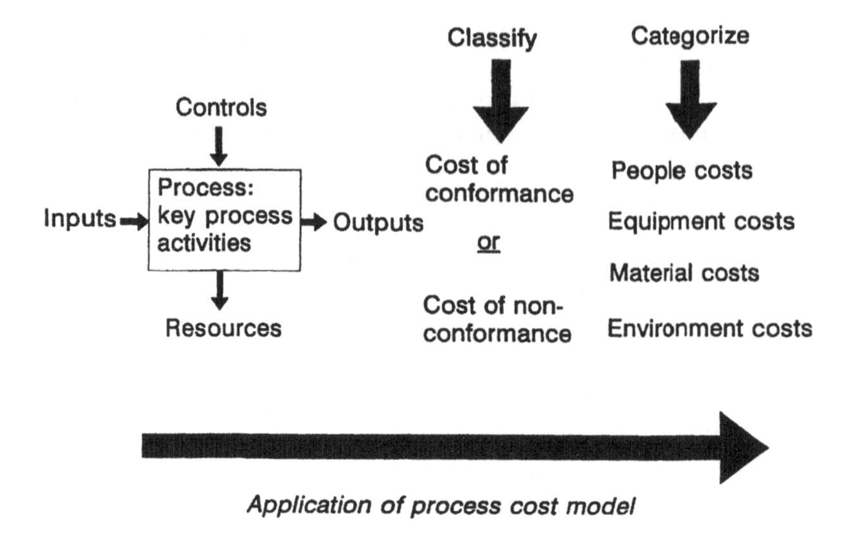

Figure 3.3 The process cost of quality model.

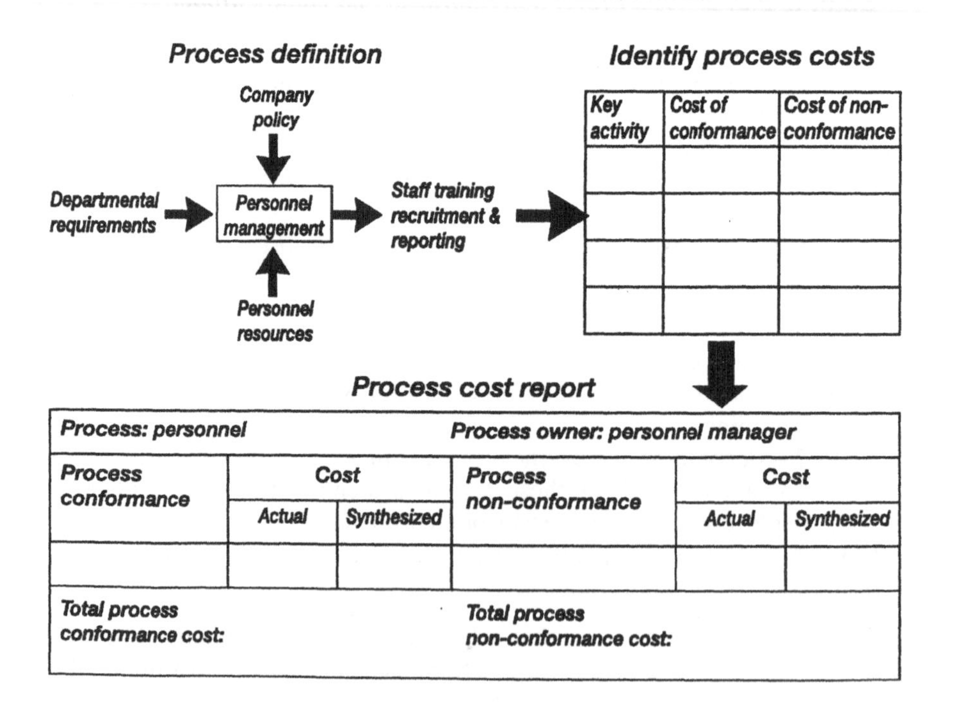

Figure 3.4 The process cost of quality model report.

3.2.2 PREVENTION, APPRAISAL AND FAILURE (PAF) COSTING

The traditional approach to reporting the cost of quality is to classify the costs as follows.

Prevention costs

Prevention costs are the costs associated with the planning and operation of a quality management system to reduce the failure and appraisal costs.

Typical prevention costs incurred by an organization include:

- **development of inspection and test equipment** in terms of the specification of such items and the associated calibration costs, but excluding the capital costs;
- **quality assurance activities** such as quality system design, implementation, audit and review (Chapter 2);
- **supplier quality assurance**, including development programmes, establishing specifications and supplier assessment;
- **training**, including both quality system specific training programmes and the more general staff development activities.

Prevention costs primarily relate to the quality assurance activities within the organization.

Appraisal costs

Appraisal costs are those costs associated with verification activities to ensure conformance to requirements although they exclude those costs associated with re-inspection following the failure of a product or service. Typical appraisal costs include:

- **incoming materials/services verification activities**, including the inspection of pre-contract trials;
- **in-process verification**, including all checking of the required parameters to ensure product or process conformance;
- **inspection equipment**, including the depreciation costs of capital equipment, maintenance costs and the cost of consumables used in the inspection activities;
- **process control activities**, including the cost of monitoring and evaluating process variation;
- **analysing, reviewing, reporting and storing** the appraisal data.

Appraisal costs primarily relate to the quality control activities within the organization.

Internal failure costs

Internal failure costs are the costs associated with the non-conformance of the product or service prior to the transfer of ownership to the customer. Typical internal failure costs include:

- **scrap** of defective product which cannot be reworked, including the material and direct/indirect labour costs associated with the rejected item;
- **rework or replacement costs** for product which does not conform to the specified requirements, including the costs associated with the identification of the remedial action and root causes;
- **modification, downgrading and concessions**, including the effort involved in obtaining agreement to reduce the specification and the loss in sales revenue associated with having to reduce the price;
- **re-inspection costs** associated with product having previously failed to meet requirement and having to be reworked.

External failure costs

External failure costs are those costs associated with non-conformance of the product and which are detected after the transfer of ownership to the customer. Typical external failure costs include:

- **rejected/returned product**, including the cost of repairing or replacing the product and the associated handling costs;
- **warranty claims and product liability** which may represent costs significantly in excess of the actual value of the product or service supplied and often include recall costs for equivalent sales to other customers;
- **customer dissatisfaction**, including the investigation of complaints, the commercial downgrading of the product or service and the potential loss of future sales.

Internal and external failure costs primarily relate to the ineffectiveness of the way in which quality is managed within the organization.

The main problem in establishing a PAF quality costing system often lies with the judgemental nature of classifying costs as either prevention, appraisal or failure and the different nature of these three concepts. Prevention represents an intention, appraisal represents an activity and failure represents an outcome. For example, a design review is an attempt to prevent errors of specification whereas the activity is one of appraisal and yet the outcome is the detection of internal failures of the designer.

Many companies fail at this classification stage in the implementation of quality costs. However, the real benefits in measuring the cost of quality are derived from analysing the ongoing trend and therefore consistency of interpretation is perhaps more important than accuracy of definition.

The main stages in the implementation of a PAF quality costing system are as follows:

1. Establish senior management support for the philosophy of measuring quality improvement through the report of quality costs.
2. Establish a cost of quality action team comprising staff from a range of functions including finance and quality assurance.
3. Identify the quality costs using the categories described above and collect data on the costs within each of the categories. Often a pilot study in one defined part of the organization is a sensible approach for a company introducing a quality cost system for the first time.
4. Collate and validate the quality cost information to ensure both the accuracy of the data and consistency in the interpretation, and establish procedures for the ongoing preparation and presentation of quality cost reports.
5. Introduce awareness training for employees to emphasize the reasons for identifying quality costs and the relationship with programmes for quality improvement in reducing the failure costs.
6. Prepare and distribute the quality cost report as a mechanism for focusing and measuring quality improvement and ensuring reaction by the recipients.

A typical format for a PAF cost of quality report is shown in Figure 3.5.

Group function:				Location:		
Current period				Year to date		
Budget	Actual	Var.	Quality costs	Budget	Actual	Var.
			Prevention costs Total prevention costs % of total costs			
			Appraisal costs Total appraisal costs % of total costs			
			Internal failure costs Total internal failure costs % of total costs			
			External failure costs Total external failure costs % of total costs			
			Total cost of quality			

Figure 3.5 Typical PAF cost of quality model report.

3.2.3 QUALITY COST REPORTING

Both the process cost approach and the PAF cost approach require the identification and categorization of cost information as illustrated in Figure 3.6.

Whilst the format of the reports may be different, the underlying philosophy remains the same. By focusing upon the cost of conformance data, the process model approach encourages improvements (reductions) in how the organization achieves quality and this is perhaps rather more consistent with an ideology of continuous improvement.

The PAF quality cost reports are traditionally employed primarily to reduce the failure costs. Upon the introduction of a quality cost system, the reporting of the information is normally presented on a relatively frequent basis (typically either monthly or quarterly), although in time annual reporting is perhaps most common.

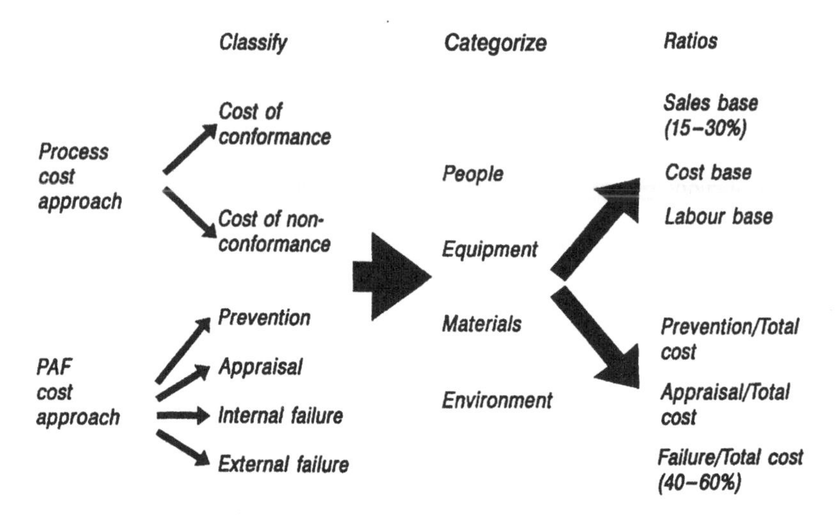

Figure 3.6 Approaches to reporting the cost of quality.

3.2.4 SUMMARY

- Quality costing systems can be established using either a process cost model or a prevention, appraisal, failure model.
- The process cost model categorizes costs as either the cost of conformance or cost of non-conformance.
- PAF costing identifies prevention, appraisal, and internal and external failure costs.
- Quality cost reports provide a detailed classification of costs and can be used to monitor progress in terms of quality development.

3.3 Quality cost performance models

3.3.1 THE TRADITIONAL VIEW OF THE COST OF QUALITY

One of the main effects of the progress made in recent years in the widespread and systematic collection and review of quality costs has been to amend management thinking on the relationship between quality and cost. The traditional view of this relationship has focused upon the concept of 'optimization', whereby at a predefined quality level, the difference between the 'returns' and the 'costs' is maximized. Fundamental to this view is the 'law of diminishing returns' which indicates that a disproportionate amount of effort is required to further improve the quality of the product or service beyond the stage of general acceptability to the customer. This traditional approach is epitomized by the view of the relationship between the cost of improving the level of quality and the return from the marketplace in terms of how much the customer is willing to pay, as illustrated in Figure 3.7.

However, when quality management is viewed as a developmental process, this traditional view has two serious flaws.

- The model assumes a certain approach to the management of quality, namely quality control rather than quality assurance or total quality management.

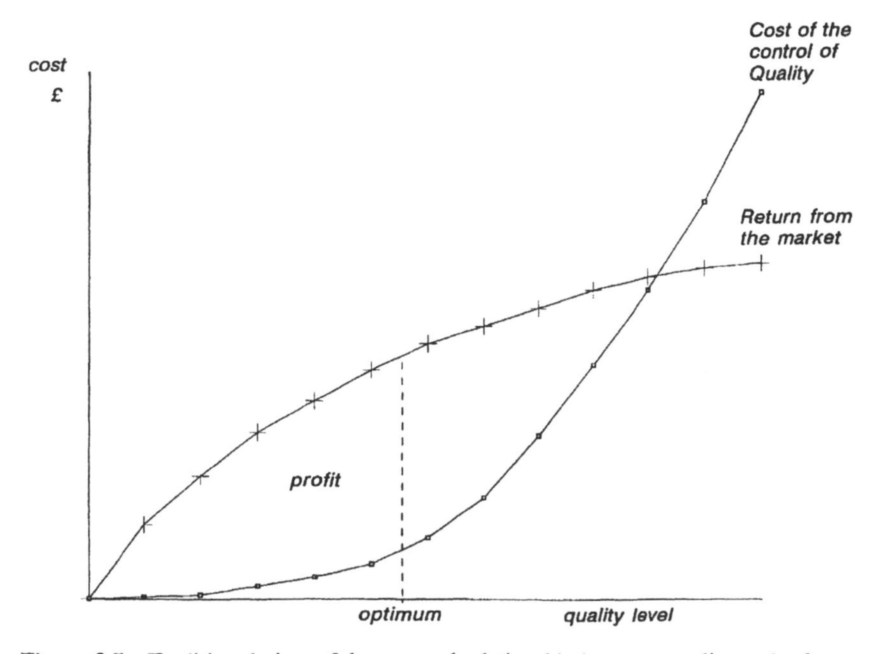

Figure 3.7 Traditional view of the external relationship between quality and value.

- The model does not take into account the potential for increasing market share through positive customer reaction to improved quality.

When traditional thinking is applied to the internal costs of managing quality the same misleading relationships emerge, whereby the optimum quality level occurs at the point at which the costs of conformance and non-conformance balance, as shown in Figure 3.8.

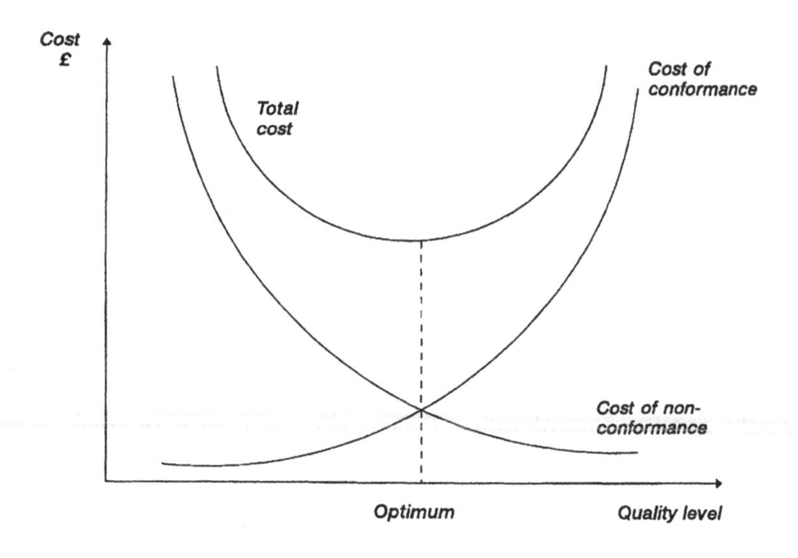

Figure 3.8 Traditional view of the internal relationship between the costs of conformance and non-conformance.

Again inherent in the misconception surrounding this view is that quality is controlled through checking/inspecting and the cost continues to rise as epitomized by the phrase 'don't spend too much money on getting the quality right'. It was the fundamental reappraisal of management thinking on the cost of quality performance models that promoted the international trends in quality development during the 1970s and 1980s.

3.3.2 THE TOTAL QUALITY VIEW OF THE COST OF QUALITY

One of the most influential corollaries of the quality development process is that mature companies actually spend less on the total cost of quality.

In terms of the PAF approach to quality costing, the relationship between the investment in prevention activities and the reduction in failure costs illustrated in Figure 3.2 suggests that as the quality level improves, the cost of achieving it reduces. This represents a powerful inducement for quality development. Similarly, using the process cost model approach, the quality improvement process involves a reduction both in the cost of non-conformance (waste and inefficiency issues are often tackled as 'bottom-up' projects involving process operators) and the cost of conformance (generally process management issues requiring 'top-down' involvement).

The investment in improvement and prevention associated with quality development leads to a relationship between cost and quality in which the satisfaction of internal and external customers reduces the overall cost of quality as shown in Figure 3.9.

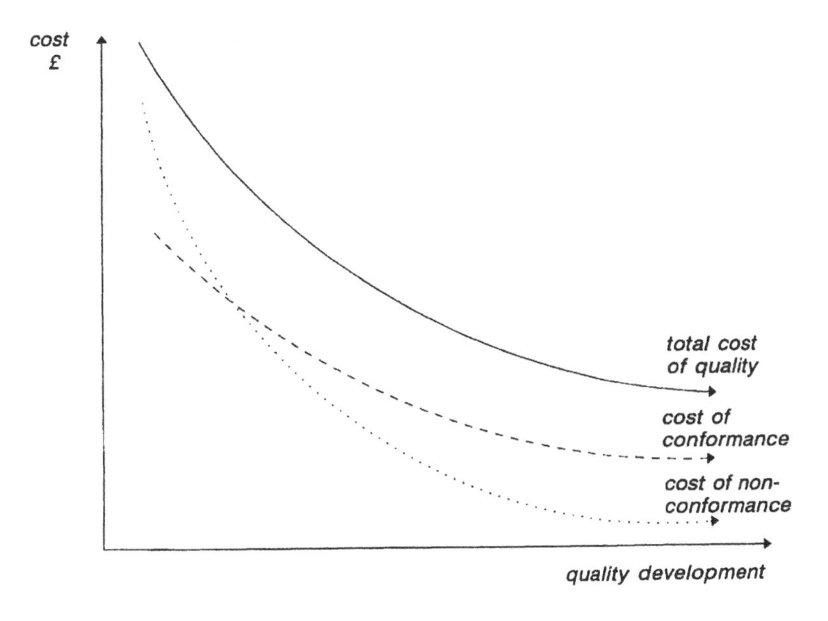

Figure 3.9 The total quality relationship between quality cost and quality development.

Certainly the empirical evidence from organizations that have measured the cost of quality over a period of time suggests that the continuous improvement quality cost performance model represented in Fig. 3.9 is a more valid approximation to reality than the migration towards an 'optimum' quality position proposed by the outdated traditional view.

3.3.3 THE TAGUCHI VIEW OF THE COST OF QUALITY

A further alternative for the relationship between the performance of cost and quality has been proposed by Taguchi in which quality is defined as: 'the loss a product causes to society after being shipped'. Whilst at first examination this may not appear to be either a very informative definition or indeed consistent with one's intuitive understanding of the meaning of the word 'quality', it does offer an interesting alternative insight into the way in which quality and cost are related. The traditional view of the cost of non-conformance is that, provided a product or service is supplied within tolerance, then the cost is zero. The cost of non-conformance, by definition, only emerges once the product or service is out of specifiction limits and therefore the cost model is a discontinuous step function as illustrated in Figure 3.10.

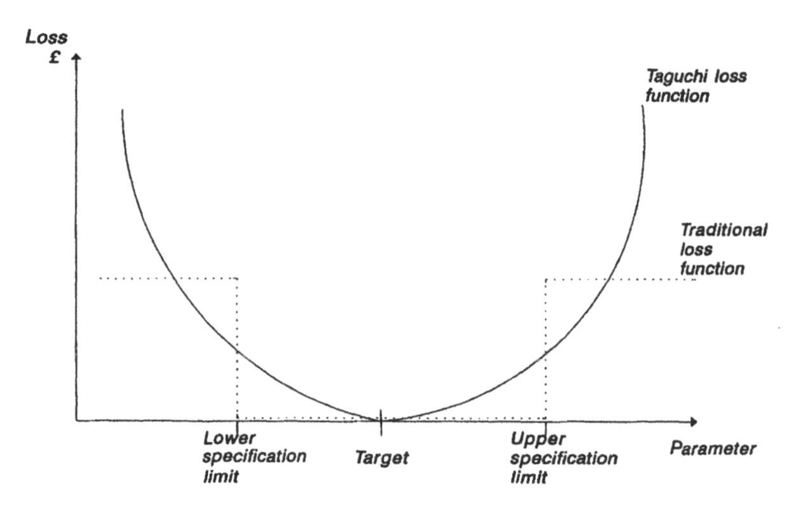

Figure 3.10 Comparison between the traditional and Taguchi views of loss due to deviation.

The Taguchi view, however, suggests that the cost or loss commences as soon as the product or service deviates from the target (or nominal) value. Further, Taguchi maintains that the relationship is quadratic and that the cost varies with the square of the deviation from the target as shown in Figure 3.10. The Taguchi loss function (see also Chapter 10) is therefore represented by:

$$L(y) = k(y - m)^2$$

loss associated with parameter y = A constant k times the square of the difference between the parameter y and the target m

Clearly the Taguchi model is a more severe case and places much greater emphasis upon providing products or services which are as close as possible to the target value and not simply within specification.

These alternative models of the relationship betwen cost and performance when the product or service lies within specification yet not at the target value begs the question: 'Which view is valid?' Certainly the Taguchi model is appropriate where:

- elemental products or services are combined to form a more complex system (for example, the assembly of a motor car, television set or railway timetabling) where the combinational effects of tolerancing can be detrimental;
- the cost of quality is to be used to establish the tolerances of the product or service.

A more complete description of the use and application of the Taguchi loss function is provided in section 10.3.

3.3.4 SUMMARY

- The traditional view of the relationship between cost and quality is that there are diminishing returns and therefore an optimum level of quality can be established.
- The total quality view suggests that as organizations develop in terms of the way in which quality is managed, then the total cost of quality reduces.
- Taguchi's loss function suggests that losses are incurred as soon as a product or service deviates from the desired value.

3.4 Benchmarking and performance measurement

3.4.1 THE ROLE OF BENCHMARKING IN QUALITY DEVELOPMENT

Another important element in measurement of quality development is the comparative analysis of performance termed **benchmarking**. Benchmarking can be defined as:

> Measuring the performance of processes within your organization, comparing these performance levels with best-in-class companies and, where deficiencies exist, using the information on best practice to improve your organization's own business processes.

The idea of comparing your performance with competitors is clearly not a new concept but has increasingly become a popular quality improvement activity due in part to the following.

- There are benefits to be gained from making performance comparisons across a range of business sectors and not simply with your competitors.

- Customers are supplied by a wide range of different types of organizations and therefore the standards for customer service may be established by other sectors of commerce or industry.
- Looking outside your own industry or business sector can lead to 'breakthrough' thinking and often the implementation of ideas from elsewhere is more readily achieved.
- Internally generated improvement objectives or targets can be unecessarily limited.

Benchmarking is particularly applicable to service organizations or departments as a mechanism for identifying quality improvements and generally the lessons learnt are very much more transferable.

3.4.2 BENCHMARKING METHODOLOGIES

The strength of using benchmarking as an aid to quality development is that it leads to comparison with different types of organization which may be at a more advanced state of development or alternatively may have different perspectives. However, this is also the potential weakness in benchmarking in that if the correct measures are not selected or valid comparisons not made, then it can become a fruitless and distracting activity.

The key to successful benchmarking is to establish the correct procedures and methodology. A typical benchmarking methodology is shown in Table 3.1.

Table 3.1 The main stages in benchmarking

Stage 1 *Identification and planning*

- Identify measures important to the customer and to the business to be benchmarked.
- Identify a benchmarking team to pilot the activity within the organization.
- Clearly define processes and current measures to be benchmarked.

Stage 2 *Data study*

- Prepare study framework for collecting comparative benchmark data.
- Identify best-practice companies and evaluate relevance of comparison.
- Arrange visits/interviews/surveys to collect data and develop understanding of best-practice processes and performances.

Stage 3 *Benchmark analysis*

- Identify critical improvement factors from the benchmark data.
- Evaluate performance gap and process improvements.
- Identify any distinguishing factors or limits to improvement and process modification.

Stage 4 *Implementation and review*

- Introduce process and performance improvements.
- Review new performance and re-evaluate performance gap.
- Review benchmarks.

3.4.3 INTEGRATED PERFORMANCE MEASURES

Based upon the old adage that 'what gets measured gets done', it is a critical dimension of quality improvement to adopt the correct measures of performance. The requirements of an integrated set of performance measurements for quality development are:

- they should reflect the business quality objectives;
- they should be coherent in the sense that measures of performance in one area (e.g. sales volume) should not conflict with measures in other areas (e.g. production efficiency);
- they should enforce the culture of customer (both internal and external) orientation;
- they are readily measureable.

The application of quality costs certainly represents a potentially integrated set of performance measures and bridges the traditional 'gap' between financial measures and operational measures which exist in so many companies.

In terms of the process of relating the measure of improvement associated with quality development to the business objectives, this is a decompositional technique whereby each objective is broken down into a series of improvement processes and hence improvement measures. An example of the decomposition of business objectives into improvement measures is shown in Figure 3.11.

Having established the performance measures to support quality development, the next stage is to specify the goals. The nature of the goals will influence the approach to improvement adopted.

- Bold goals promote the re-engineering of the process.
- Limited goals promote incremental improvement in the process.

The application of performance measures and the selection of goals for the quality development process are discussed further in Chapter 5.

3.4.4 SUMMARY

- Benchmarking can contribute to the quality development process by providing insights into new approaches and new levels of performance.
- Benchmarking needs to be applied in a structured manner to ensure consistency of interpretation and comparison.
- The performance measures used to support quality development need to be integrated and coherent.

First level planning

Business objective	Improvement objective	Performance measure	Goal	Priority	Department	Team leader
Improved return on capital employed	Improved throughput per person	Kg per direct employee	1000 Kg per employee	1	Manufacture/ Engineering	
Improved sales delivery performance	Improved sales forecast accuracy	Actual versus forecast percentage	95% accuracy	2	Sales/Marketing	
Improved new product development	Reduced development lead times	Time taken to introduce new products	26 weeks	3	NPD/Manufacture	

Second level planning

Improvement objective	Critical processes	Performance measure	Goal	Priority	Department	Team leader
Improved throughput per person	Process waste	Line output raw material usage	>5% for all multi-stage	1	Manufacturing	
	Set-up times	Changeover times	<15 minutes	2	Manufacturing/ Technical	
	Maintenance downtime	Lost time for processes	<5% of available process time	3	Engineering	

Figure 3.11 Decomposition of business objectives into integrated performance measures.

4 Motivation for quality

4.1 Developing a 'quality culture'

4.1.1 MOTIVATION AND QUALITY DEVELOPMENT

As described in Chapter 1, the quality development of any organization involves systems, techniques and people. Perhaps the most critical and yet most difficult of these three dimensions to develop are the people. The development of individuals and teams is essential to the quality development of any organization. Developing quality systems and quality techniques are relatively straightforward management challenges when compared to managing the cultural change associated with motivating people for quality. The dimensions of systems and techniques are rich with methods and guidelines; however, progress in the third dimension, people, is much less mechanistic or prescriptive and in general is much slower. Many organizations falter on the road to quality development because insufficient effort or priority is given to improving the approach and attitude of the people involved.

In terms of the stages of quality development described in Chapter 1, the extent of cultural (people) change increases as organizations move from a systems orientation through an improvement orientation to a prevention orientation. If two of the key elements of cultural change, teamwork and management style, are mapped onto this developmental framework it can be seen that:

- **teamwork** evolves from having a very small role within the systems orientation stage to being an important but separate activity during the improvement orientation to being an integral part of the organization in the prevention orientation stage;
- **management style** evolves from awareness to involvement to commitment.

These stages of cultural development are illustrated in Figure 4.1.

An integral part of the total quality approach is the harnessing of the skills of the people within the organization. When considering the assets of any organization (equipment, buildings, products, knowledge, capital, etc.) the only asset which is creative, which can take the company forward, is the

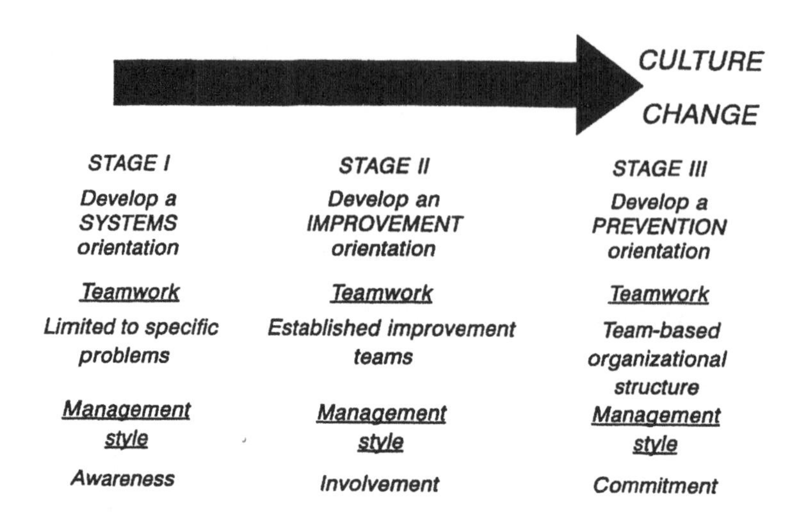

Figure 4.1 The cultural changes during quality development.

people. In many immature organizations this latent talent of the people within is ignored as they are seen, in the words of Henry Ford, as 'hands not minds'. Very often individuals who are seen as employees capable of only mundane tasks within the workplace are involved in creative, organizational activities outside the work environment such as wardens in the local church or managers of local sports teams. People have to be part of the quality development of an organization. Organizations develop at the rate at which their people develop.

Central to the organizational development associated with quality improvement is 'culture' change. Organizational culture is an important concept when seeking to promote quality development although culture is difficult to precisely define or to measure. A simple definition of the culture of an organization is:

> The shared values and norms of behaviour of the individuals within the organization.

If, for example, the managers within a company promote an attitude of 'let's try to cheat our customers by getting away with as much sub-standard product as possible' then it should come as no surprise if the employees adopt this same attitude with the company. If the organization blames and criticizes individuals for mistakes or errors then the people will respond by being deceitful and evasive. If, however, people are trusted and respected then this will be reflected in their behaviour and values.

For people to produce quality work they must be motivated and care about the work they do. To be motivated individuals must have a motive and this sense of purpose must be clearly communicated to people. Motivating individuals and creating a culture change represents perhaps the most significant management challenge on the road to quality development.

4.1.2 PEOPLE AND MOTIVATION

A great deal of the writing on management science during the twentieth century has been devoted to the relationship between people and work. Understanding the answer to the question of 'what motivates people?' is fundamental to managing the cultural change required for quality development. Among the more significant developments in understanding the relationship between people and work have been the contributions from:

- F.W. Taylor;
- A.H. Maslow;
- F. Herzberg;
- D. McGregor.

F.W. Taylor

The contribution of F.W. Taylor to management thinking in industry during the first half of the twentieth century was significant. Taylor advocated a 'scientific' approach to management where the emphasis should be placed upon control and methods. This management approach was used extensively in the USA during this period and was attributed with the creation of a strong American industrial base after the Second World War. In many ways the style of management promoted by Taylor was appropriate to a situation where organizations were employing large, mainly immigrant workforces. Work measurement, method study and bonus schemes were very much part of this scientific approach to management, but increasingly from the 1970s onwards these techniques were seen to be working against the development of a quality culture. Control and payment-based incentives were not delivering the quality products and service that were increasingly coming out of Japan where managers had largely rejected the Taylor approach.

A.H. Maslow

A.H. Maslow in his work on people and motivation identified a hierarchy of human needs. Maslow suggested that as individuals developed and achieved the needs associated with one level then they would seek the needs at the subsequent (higher) levels. This model of human motivation, although proposed in the 1940s, is still relevant to quality management issues of the 1990s. As organizations develop and mature then the mechanisms for motivating people become more complex and abstract. Maslow's hierarchy of needs is illustrated in Figure 4.2 and also offers an important insight into the problems of international competition for developed nations where competitors in less developed (often low-wage economies) can achieve levels of motivation through the provision of rather more basic human needs.

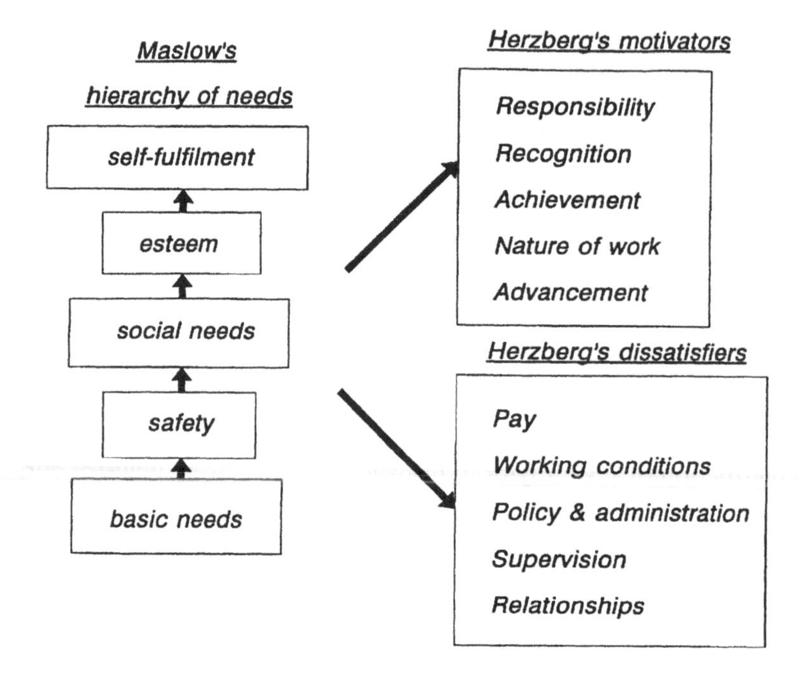

Figure 4.2 Factors affecting motivation.

F. Herzberg

F. Herzberg further developed the ideas relating motivation in the workplace to needs by classifying factors as either satisfiers or dissatisfiers. The factors affecting dissatisfaction were termed the 'hygiene' factors and included:

- salary;
- working conditions;
- policy and administration;
- supervision arrangements;
- relationships.

These factors were seen by Herzberg as potential causes of dissatisfaction but in themselves could not provide long-term satisfaction. In other words individuals could only become demotivated by these factors (dissatisfiers) and never become sustainably motivated.

The factors affecting job satisfaction identified by Herzberg included:

- responsibility;
- recognition;
- achievement;
- nature of the work;
- advancement.

These factors were seen to produce satisfaction and were termed the 'motivators' because they could produce a positive attitude to work and often for a longer duration. This important distinction between concentrating on factors which have, at best, a neutral effect upon motivation and focusing upon more positive motivational factors such as responsibility is an essential management understanding. The cultural change in management style associated with quality development is very much concerned with the focus upon the positive 'motivators'. The experiences of organizations which are successful at delivering quality improvement has emphasized the need to manage both the hygiene factors and the motivator factors. Quality development that focuses only upon motivators will be naive. Similarly focusing purely on the hygiene factors will not produce long-term benefit. Both groups of factors need improvement in order to provide sustainable quality development.

D. McGregor

D. McGregor proposed two opposing behavioural models in terms of the way in which people approach work. The first McGregor described as Theory X, which states that humans:

- inherently dislike work and will avoid it if possible;
- need to be controlled and coerced to provide adequate work effort;
- wish to avoid responsibility and have little ambition above satisfying basic needs.

The alternative model was described by McGregor as Theory Y in which humans:

- see work as a natural activity as important as rest or play;
- exercise self-control and are committed to accepted objectives;
- seek responsibility and possess creativity which is largely untapped.

These opposing behavioural viewpoints are fundamental to the management perspective in dealing with motivation. Are people Theory X or Theory Y? The significance of McGregor's work is the link made between the way in which people are managed and the way in which they behave. If people are treated as Theory X then they will behave as Theory X. If, on the other hand, the management culture promotes a Theory Y viewpoint, then people are more likely to behave accordingly.

This maturing view as to how people should be managed and motivated is reflected in the developments in quality management over the past 30 years. Increasingly the focus in recent years has been upon the involvement, empowerment and motivation of the workforce. The systems and techniques of quality management will only be implemented effectively if the people believe in the process and support the objectives.

In service industries, quality starts and ends with the people. In manufacturing companies, despite the advances in technology and automation, customer satisfaction is still largely delivered by the people.

4.1.3 MANAGEMENT STYLE

As with 'culture' the phrase 'management style' is an often used descriptor of the relationship between people and motivation and is generally a complex concept that is difficult to quantify. In simple terms management style can be defined as:

> The collective methods and approaches used to motivate people within the organization.

The complexity of this concept lies in attempting to identify the subtle combination of explicit and implicit activities employed by managers within an organization to achieve the objectives. Very often the management style and values are not implicitly communicated and are therefore passed down through the management structure explicitly. For example, a young manager who joins an organization will observe the way in which more senior colleagues behave and will also develop a set of criteria in terms of management approach that coincides with progression and promotion within the organization. If the senior managers in a company exhibit a control-orientated management approach based upon Theory X then this will promote this style of management among middle and junior managers and will inhibit cultural change (why should senior managers change a personal style of management that has brought them individual success in terms of promotion?).

To develop a quality-orientated approach firstly senior managers need to recognize the existing style of management, identify a management approach which supports quality development and finally communicate the required culture in a credible manner to the organization. This means adopting an appropriate management style rather than simply expounding one. People listen to what managers say, but they also watch what they do!

Developing a quality culture is a senior management responsibility and involves:

- identifying a style of management to support quality development;
- effective communication.

There is of course no single style of management upon which to base quality development. There are, however, certain features of the approach to the management of people which support total quality and these include:

- adopting an 'enabling' rather than 'controlling' style of management (the manager's viewpoint should be 'how can I help my people do their work better' rather than 'how can they help me to do my work');
- management values based upon respect for individuals for honesty and for integrity.

From this style of management derives the role of the manager in the organization who should be involved in the support and facilitation (rather than control) of subordinates. This role is illustrated in an alternative viewpoint of the 'organogram' shown in Figure 4.3 whereby the managers support rather than attempt to control the efforts of others.

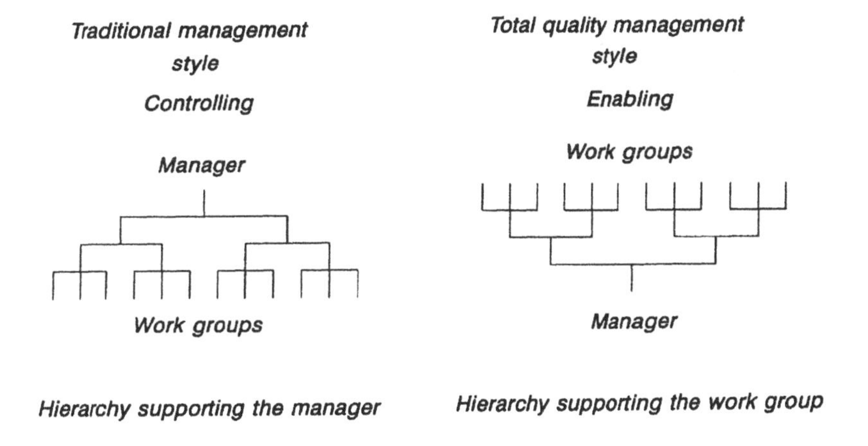

Figure 4.3 The manager's role in the organization.

The second key element of management style concerns communication. In organizations arranged in functional 'chimneys' (where work is arranged on a functional rather than process basis) the communications are traditionally poor. Poor organizational communication inevitably leads to errors and hence to quality problems. Typical inhibitors to good communication within organizations include:

- lack of opportunity;
- fear of reprisal;

- people too busy;
- lack of priority;
- departmental boundaries.

Good communications are critical to good quality and good customer service and certainly as companies mature and develop an improvement orientation, a greater priority and emphasis is given to internal and external communication and to communication being a two-way process. Techniques such as briefing sessions or 'talk-back' sessions in which managers schedule time specifically for communicating are commonplace in organizations at the second stage of quality development (as described in Chapter 1). As companies mature yet further then reorganization often takes place to eliminate departmental boundaries through the use of teams arranged according to business processes.

4.1.4 SUMMARY

- The motivation of people is a key ingredient in quality development.
- Models of human behaviour and motivation emphasize the need to address both the basic needs and also the higher level 'motivators' such as responsibility and recognition.
- A quality culture requires an enabling management style and effective internal and external communications.

4.2 Teambuilding and the role of teamwork

4.2.1 TEAMWORK AND QUALITY DEVELOPMENT

Research into the approaches used to implement quality development has identified many different methodologies for example:

- techniques led;
- 'guru' philosophy led;
- customer led;
- process led.

The common factor, however, is that all organizations on the road to quality development use teams. **Teamwork** is universally adopted as the vehicle for change and the organizational mechanism for involving people in quality improvement. The benefits of a team-based approach include:

- improved solutions to quality problems;
- improved ownership of solutions;
- improved communications;
- improved integration.

Many of the quality improvement tools and techniques explicitly require team-based application and benefit from the range of skills and insights provided by teamwork. Perhaps the most significant organizational change associated with quality development is the increased emphasis upon teamworking. The 'quality improvement' stage of development described in Chapter 1 is very much devoted to the operation of improvement teams. Prevention-orientated companies integrate these team-based activities into the organizational framework. Rather than teamworking being a separate activity associated only with quality improvement projects it becomes part of normal operations within the workplace.

During this adoption of a more team-based culture, it is important for organizations to correctly understand the role of teamwork. It is incorrect, for example, to assume all activities or improvements need to be generated or derived from teams. Individuals are obviously still an important component of the quality developed organization – it is simply the balance between individual and team-based activity that is changed. By harnessing the benefits of teamwork, companies combine powerful techniques with improved motivation.

4.2.2 BUILDING EFFECTIVE TEAMS

The words 'teams' and 'teamwork' are used widely throughout organizations to describe groups of people and group activity. This widespread use of the descriptor 'team' has led to the concept of teamworking being taken rather for granted. Individuals may well be organized into identifiable groups but they may not be working as a team. Given that teamworking underpins much of the cultural change and the application of the techniques of quality development, then it is important to ensure that teambuilding is carried out effectively. For example, if a quality improvement team is to be assembled, who should be included in the team? Basic guidelines for building effective teams suggest that team members should include:

- those individuals affected by the problem;
- those individuals likely to be involved;
- those individuals with appropriate skills.

This responsibility or skills dimension to teambuilding is clearly important as all teams have a basic knowledge requirement and should involve those people who understand the objectives and will own the solutions. This functional viewpoint is, however, incomplete when attempting to build effective teams. The dynamics and balance of the team are equally, if not more, important to the eventual effective operation of any group of individuals. The work of Belbin at the Henley Management Centre described below in section 4.2.3 identified that:

- balanced teams usually succeed;

- unbalanced teams usually fail.

The balance of a team is achieved through an appropriate collection of individuals having different yet complementary team roles. As described below all individuals exhibit certain identifiable team roles and effective teams are built through balancing the combination of these team roles. The assumption that any group of individuals with the required skills can be brought together is naive and unlikely to succeed.

The climate in which the team operates is also important and this requires:

- a supportive atmosphere where individuals are not fearful of contributing;
- a welcoming atmosphere where individual contributions are encouraged;
- mutual respect;
- no one view or individual dominating;
- opposing views to be encouraged and handled positively.

Recognizing the significance of team roles and team balance and the importance of providing the correct climate are important prerequisites for enabling quality development through teamwork.

4.2.3 TEAM ROLES

Different individuals have different contributions to make to the operation of a team. Much of the benefit of teamworking comes from the synthesis of different types of individual input. These types of individual contribution to the working of a team were grouped into certain identifiable team roles by Belbin. Belbin found by examining the performance of different team groupings that individuals working within a team behave in accordance with one of eight basic team roles (later the number of identifiable roles was increased).

To identify an individual's team role an analysis technique based upon a self-perception inventory was developed and this is shown in Figure 4.4. For each of the seven sections of the self-perception inventory there are ten points to be distributed according to the individual's own view of his or her working within a team. These ten points can be relatively evenly distributed over the eight alternative statements in each section or may be awarded to only one or two statements. The important aspect of the analysis is that the individual should apportion the ten points available in such a way as to reflect the extent to which each statement accurately reflects one's view of oneself.

Having completed each of the seven sections of the self-perception inventory, the results should be summarized on the points table and finally onto the analysis sheet as shown in Figures 4.5 and 4.6.

Section I: What I believe I can contribute to a team:

	points	yes
(a) I think I can quickly see and take advantage of new opportunities.		
(b) I can work well with a very wide range of people.		
(c) Producing ideas is one of my natural assets.		
(d) My ability rests in being able to draw people out whenever I detect they have something of value to contribute to group objectives.		
(e) My capacity to follow through has much to do with my personal effectiveness.		
(f) I am ready to face temporary unpopularity if it leads to worthwhile results in the end.		
(g) I am quick to sense what is likely to work in a situation with which I am familiar.		
(h) I can offer a reasoned case for alternative courses of action without introducing bias or prejudice.		

Section II: If I have a possible shortcoming in teamwork, it could be that:

	points	yes
(a) I am not at ease unless meetings are well-structured and controlled and generally well-conducted.		
(b) I am inclined to be too generous towards others who have a valid viewpoint that has not been given a proper airing.		
(c) I have a tendency to talk a lot once the group gets on to new ideas.		
(d) My objective outlook makes it difficult for me to join in readily and enthusiastically with colleagues.		
(e) I am sometimes seen as forceful and authoritarian if there is a need to get something done.		
(f) I find it difficult to lead from the front, perhaps because I am over-responsive to group atmosphere.		
(g) I am apt to get too caught up in ideas that occur to me and so lose track of what is happening.		
(h) My colleagues tend to see me as worrying unnecessarily over detail and the possibility that things may go wrong.		

Section III: When involved in a project with other people:

	points	yes
(a) I have an aptitude for influencing people without pressurizing them.		
(b) My general vigilance prevents careless mistakes and omissions from being made.		
(c) I am ready to press for action to make sure that the meeting does not waste time or lose sight of the main objective.		
(d) I can be counted on to contribute something original.		

continued. . .

(e) I am always ready to back a good suggestion in the common interest.
(f) I am keen to look for the latest in new ideas and developments.
(g) I believe my capacity for cool judgement is appreciated by others.
(h) I can be relied upon to see that all essential work is organized.

Section IV: My characteristic approach to group work is that:

	points	yes
(a) I have a quiet interest in getting to know colleagues better.		
(b) I am not reluctant to challenge the views of others or to hold a minority view myself.		
(c) I can usually find a line of argument to refute unsound propositions.		
(d) I think I have a talent for making things work once a plan has to be put into operation.		
(e) I have a tendency to avoid the obvious and to come out with the unexpected.		
(f) I bring a touch of perfectionism to any team job I undertake.		
(g) I am ready to make use of contacts outside the group itself.		
(h) While I am interested in all views I have no hesitation in making up my mind once a decision has to be made.		

Section V: I gain satisfaction in a job because:

	points	yes
(a) I enjoy analysing situations and weighing up all the possible choices.		
(b) I am interested in finding practical solutions to problems.		
(c) I like to feel I am fostering good working relationships.		
(d) I can have a strong influence on decisions.		
(e) I can meet people who may have something new to offer.		
(f) I can get people to agree on a necessary course of action.		
(g) I feel in my element where I can give a task my full attention.		
(h) I like to find a field that stretches my imagination.		

Section VI: If I am suddenly given a difficult task with limited time and unfamiliar people:

	points	yes
(a) I would feel like retiring to a corner to devise a way out of the impasse before developing a line.		
(b) I would be ready to work with the person who showed the most positive approach, however difficult he might be.		

continued. . .

	points	yes
(c) I would find some way of reducing the size of the task by establishing what different individuals might best contribute.		
(d) My natural sense of urgency would help to ensure that we did not fall behind schedule.		
(e) I believe I would keep cool and maintain my capacity to think straight.		
(f) I would retain a steadiness of purpose in spite of the pressures.		
(g) I would be prepared to take a positive lead if I felt the group was making no progress.		
(h) I would open up discussions with a view to stimulating new thoughts and getting something moving.		

Section VII: With reference to the problems to which I am subject in working in groups:

	points	yes
(a) I am apt to show my impatience with those who are obstructing progress.		
(b) Others may criticize me for being too analytical and insufficiently intuitive.		
(c) My desire to ensure that work is properly done can hold up proceedings.		
(d) I tend to get bored rather easily and rely on one or two stimulating members to spark me off.		
(e) I find it difficult to get started unless the goals are clear.		
(f) I am sometimes poor at explaining and clarifying complex points that occur to me.		
(g) I am conscious of demanding from others the things I cannot do myself.		
(h) I hesitate to get my points across when I run up against real opposition.		

Figure 4.4 Belbin's self-perception inventory.

The points awarded in each of the columns of the analysis sheet are then totalled. The highest score represents the **primary** team role and the second highest score represents the **secondary** team role. Where scores are equal then the individual can exhibit both the team roles when working within a team. The columns shown on the analysis sheet correspond to the eight basic team roles identified by Belbin which are:

- company worker;
- chairman;
- shaper;
- plant;
- resource investigator;

item / section	a	b	c	d	e	f	g	h
I								
II								
III								
IV								
V								
VI								
VII								

Figure 4.5 Points table for self-perception inventory.

SECTION	CW	CH	SH	PL	RI	ME	TW	CF
I	g	d	f	c	a	h	b	e
II	a	b	e	g	c	d	f	h
III	h	a	c	d	f	g	e	b
IV	d	h	b	e	g	c	a	f
V	b	f	d	h	e	a	c	g
VI	f	c	g	a	h	e	b	d
VII	e	g	a	f	d	b	h	c
TOTAL								

Figure 4.6 Self-perception inventory analysis sheet.

- monitor-evaluator;
- team worker;
- completer-finisher.

A brief profile for each of these team roles is given below in Figure 4.7 together with their positive qualities and allowable weaknesses.

Company Worker

Typical Features:
Conservative, dutiful, predictable.

Positive Qualities:
Ability to organize and work hard, practical common sense, self-discipline.

Allowable Weaknesses:
Lack of flexibility, unresponsiveness to unproven ideas.

Chairman

Typical Features:
Calm, self-confident, controlled.

Positive Qualities:
Capacity for treating and welcoming all potential contributors on their merits and without prejudice. Strong sense of objectives.

Allowable Weaknesses:
No more than ordinary in terms of intellect or creative ability.

Shaper

Typical Features:
Highly strung, outgoing, dynamic.

Positive Qualities:
Drive and a readiness to challenge inertia, ineffectiveness, complacency and self-deception.

Allowable Weaknesses:
Proneness to provocation, irritation and impatience.

Plant

Typical Features:
Individualistic, serious-minded and unorthodox.

Positive Qualities:
Genius, imagination, intellect and knowledge.

Allowable Weaknesses:
Up in the clouds; inclination to disregard practical details and protocol.

Resource Investigator

Typical Features:
Extroverted, enthusiastic, curious and communicative.

Positive Qualities:
Capacity for contacting people and exploring anything new. Ability to respond to challenge.

Allowable Weaknesses:
Tendency to lose interest once the initial fascination has passed.

Monitor-Evaluator

Typical Features:
Sober, unemotional and prudent.

Positive Qualities:
Judgement, discretion and hard-headedness.

Allowable Weaknesses:
Lack of inspiration and the ability to motivate others.

Team Worker

Typical Features:
Socially orientated, rather mild, sensitive.

Positive Qualities:
Ability to respond to people and situations and to promote team spirit.

Allowable Weaknesses:
Indecisiveness at moments of crisis.

Completer-Finisher

Typical Features:
Painstaking, orderly, conscientious and anxious.

Positive Qualities:
Capacity for follow-through and perfectionism.

Allowable Weaknesses:
Tendency to worry about small things, reluctance to let go.

Figure 4.7 Individual profiles of Belbin's team roles.

The real significance in identifying/classifying an individual's team roles is that it brings a degree of method to the formation of teams. By understanding these alternative team contributions then more balanced groups of individuals can be selected. Also, providing the team members with an understanding of their own role in the team and the role of fellow team members makes people more appreciative of the contributions of others. Once it is understood that a certain team member is, say, a 'team worker' by nature, there is no benefit to be gained by giving that individual team responsibilities that require strong motivational or decision-making abilities. Similarly, a team that has recognized that it does not possess a naturally creative team member (a 'plant' for example) can compensate by allocating proportionally more of its time and effort to the creative requirements of the project (for example by brainstorming).

As with many industrial problems, the problem of building effective teams is lessened when an appropriate technique (in this case team role profiling) for analysing the problem is provided. Teams can be more adaptive and function more coherently once the team members understand their own and each other's role in the process. From a team dynamic point of view, therefore, a well balanced team would normally comprise:

- a good leader;
- a creative member;
- a range of team players;
- contributions from team workers, company workers and completer-finishers;
- flexibility.

Belbin's analytical approach is not the only technique for improving the effectiveness of teambuilding. It does, however, provide an indication of the benefits of using a systematic approach to building teams. When these criteria of teambuilding, interaction and dynamics are added to the skills and knowledge requirements of the members then more effective teams result.

4.2.4 SUMMARY

- Teamworking is a critical factor in the quality development process and is a requirement of many of the advanced tools and techniques.
- Building effective teams requires individuals who provide the necessary knowledge or skills but also combine together to produce balanced team dynamics.
- Techniques such as Belbin analysis to identify team roles assist in the selection of more balanced teams.

4.3 Types of quality improvement teams

4.3.1 WINNING TEAMS

As described above in section 4.2 the composition of the team in terms of a balance of team roles is one of the important considerations in building successful quality improvement teams. Many organizations maintain a record of the primary and secondary team roles of individuals and use the type of matrix shown in Figure 4.8 to create and evaluate team composition.

A balanced team does not necessarily contain every one of the individual roles but instead should contain a reasonable spread of team roles. Improvement teams, for example, that are predominently 'shapers' or 'company

Team composition

List the names of several of the people (preferably subordinates) with whom you work.

Tick the team role which best appears to fit each one.

Is your team balanced?

Name	*Company Worker CW*	*Chairman CH*	*Shaper SH*	*Plant PL*	*Resource Investigator RI*	*Monitor-Evaluator ME*	*Team Worker TW*	*Completer-Finisher CF*
1.								
2.								
3.								
4.								
5.								
6.								

Figure 4.8 Team composition matrix.

workers' will often lack the cohesion or focus (respectively) to operate effectively. Well balanced teams are generally also more flexible in dealing with changing situations which may occur during the course of the improvement project.

Another important element of team effectiveness is team size. Many organizations in the early stages of developing an improvement orientation discover that the size of the team is critical for two reasons:

- individuals are not used to working in teams and therefore find the 'large' team format intimidating;
- team facilitation skills are generally undeveloped within the organization.

For these reasons many companies have more success with smaller teams of four or five people than they do with larger teams of over, say, eight people. There is clearly a balance between having smaller, more manageable teams and larger, more balanced teams and often the balancing mechanism is team facilitation as described below in section 4.4. As organizations mature and teamworking becomes widespread, the criticality of the team size and the need for external facilitation become less.

Another important consideration in the building of winning teams is the nature of the team, whether the team is a deployed cross-functional group or a voluntary functionally based group.

4.3.2 DEPLOYED, CROSS-FUNCTIONAL TEAMS

As described in Chapter 8 there are basically two forms of quality improvement project. One form of improvement team tackles projects deployed from the organization's quality objectives. This 'top-down' improvement team usually comprises designated individuals from different functions (departments) tackling defined quality problems over a specific period of time.

The effective configuring of deployed improvement teams (sometimes referred to as **corrective action teams**) requires individuals to be selected on the basis of:

- their knowledge of both the problem (and its context) and the potential solutions, and also their ownership of the problem;
- their ability to function effectively as a balanced team.

Critical factors in the success of such improvement teams include:

- good facilitation;
- creative input;
- completer-finisher skills.

Very often team operation will also benefit from the contribution of 'outsiders' who are individuals with appropriate analytical or team skills but not necessarily directly involved with the problem.

4.3.3 FUNCTIONAL IMPROVEMENT TEAMS

The second general form of quality improvement team is the 'bottom-up' approach whereby individuals within a functional area volunteer to form an improvement group (sometimes referred to as **quality circles** or **quality improvement teams**). Such improvement teams generally identify and prioritize their own improvement objectives and projects. Functional improvement teams operate on an ongoing basis whereby the group progresses from one project to the next, often with the composition of the group changing.

The scope for pre-selecting a balanced team is often more restricted in the case of functionally based teams and the critical success factors include:

- good team leadership and training;
- well structured meetings;
- management support.

As discussed in Chapter 8, it is important that a well motivated, enthusiastic improvement team has the requisite problem-solving skills for the task in hand. The training of the team members is therefore extremely important and all the members need to be competent in the various quality improvement techniques in order to contribute to the team effort. In addition to these problems-solving skills, team effectiveness also benefits if the members have been trained in teamworking skills, including:

- an understanding of team roles;
- team facilitation;
- team responsibilities.

Team meetings need to be well planned to ensure the availability of individuals and well structured with some form of systematic methodology (as described in Chapter 8). Clear agendas for team meetings improve the effectiveness of team focus and the duration of the meetings should be defined in advance. An ethos of arriving at meetings on time (in respect to other team members) and ensuring that delegated tasks (data collection, trials, information gathering) are completed in time for the team meetings is also important.

Finally perhaps the most common reason for the undermining of team-based quality improvement efforts is lack of management support. Middle managers in particular can feel 'threatened' by the improvement activities of functionally based teams which could be seen as usurping the responsibilities of the manager. It is important therefore that managers at all levels are comfortable with, and supportive of, the quality improvement efforts.

4.3.4 SUMMARY

- Team balance and team size are critical factors in the effectiveness of quality improvement teams.
- Deployed teams require individuals who will contribute to both the understanding of and solution to the problem and also to the balance of the team.
- Functionally based improvement teams, in addition to balance, require a framework of training, management support and a practical improvement methodology.

4.4 Team leadership and facilitation

4.4.1 TEAM LEADERSHIP

As with many areas of management, the leadership of the team is critical to the effectiveness of the team. The main roles of the team leader within the improvement team are illustrated in Figure 4.9.

The team leader's role is therefore primarily concerned with the operation of the **process**. The team leader should not dominate the group but should provide impetus to the project through the efforts of the team members.

The question arises, however, as to who should be the team leader? Possible alternatives to the selection of the team leader include:

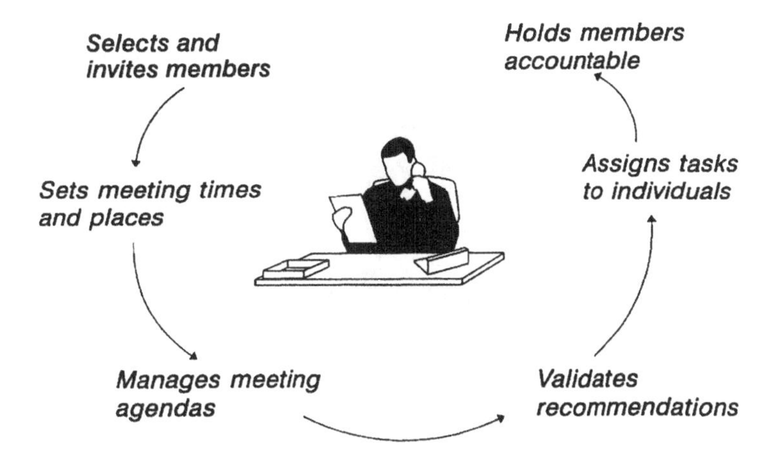

Figure 4.9 The role of the team leader.

- the most 'senior' person in the team;
- the person most directly involved in the problem;
- the person with the most appropriate team role;
- election by the other team members.

Selection of the most senior manager as the team leader is not necessarily the most appropriate choice. Very often a senior manager within a team will simply reinforce existing organizational structures and may inhibit team contribution. Certainly a strong sense of ownership of the problem coupled with an appropriate team profile (for example 'chairman' or 'shaper') generally is a more successful recipe for effective team leadership. The problem with allowing the team members to select the leader is that this does not enable the team leader to select the members. With cross-functional improvement teams it is normal to designate the leader who would then be involved in the selection of the team members. In functionally based teams it is more common to seek consensus for the appointment of the team leader.

4.4.2 TEAM FACILITATION

Team facilitation and the role within the organization of facilitators is, unlike leadership, very often a new concept to many companies. The problem is, however, that facilitation is often most important when organizations first begin to develop an orientation of improvement through teamwork. The main roles of the facilitator within the improvement team are illustrated in Figure 4.10.

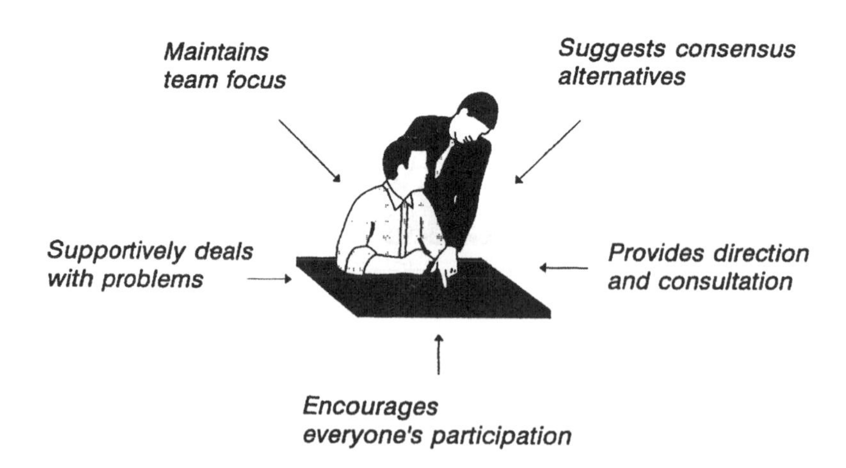

Figure 4.10 The role of the team facilitator.

The facilitator's role is therefore primarily concerned with the **team dynamics**. The facilitator should provide a cohesive influence and ensure that all the team members are providing an effective input. In addition facilitators are usually experienced in the problem-solving techniques required by the team to bring about improvement although very often the facilitator will have little direct knowledge of the problem at hand. Typically 'team workers' are effective facilitators and, as with many of the aspects of quality development, good facilitation skills require training.

As organizations develop, the role of the manager or team leader becomes much more concerned with facilitation (an enabling management style) rather than control. Where the team size is small, typically four or five team members, then it is possible for the team leader to combine the role of leader and facilitator. Recognizing the difference between these team roles and understanding the need for both to be carried out to ensure team effectiveness is essential to quality improvement team operation.

4.4.3 TEAM DEVELOPMENT AND MONITORING CULTURE CHANGE

A simple measure of the quality development of an organization is the extent to which teamwork is adopted and effectively applied. The need to create balanced quality improvement teams requires an organization to recruit and nurture a range of team players, yet typically organizations possess a limited, polarized set of such players. Figure 4.11 illustrates the findings of research into the primary team roles of employees of a major

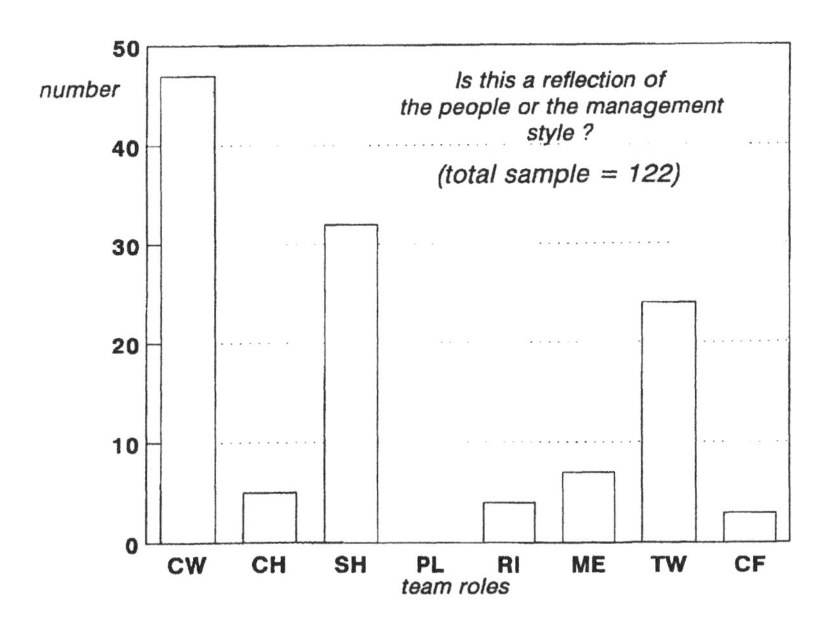

Figure 4.11 The distribution of primary team roles within a UK manufacturing company.

international manufacturing organization and indicates that three team roles dominate which mitigates against building effective teams.

This type of company-wide analysis of team roles is one simple measure of the 'culture' within an organization. Teamworking is only one dimension of the cultural change needed for quality development. Self-perception audits of the distribution of team roles within an organization is a simple mechanism for measuring culture change. As individuals experience teamworking and mature in terms of their contribution to the operation of a team, the self-perception of team functioning can change and the organization can develop a more balanced profile over, say, a three to five year period.

4.4.4 SUMMARY

- Team leadership is an important role to ensure the team organization and operation is managed effectively.
- Team facilitation ensures the team dynamics are managed effectively.
- The distribution of team roles can be used as a measure of the culture change needed for quality development.

5 Total quality management

5.1 Developing continuous improvement through total quality

5.1.1 THE IMPORTANCE OF QUALITY DEVELOPMENT

So what is total quality management? The acronym 'TQM' is perhaps one of the most widely used business expressions and has certainly been liberally used as a generic descriptor of the quality development efforts in recent years. In an academic sense, total quality is the integrated application of all the techniques described throughout this book within some kind of coherent implementation framework. In more simple language total quality management can be thought of as 'involving everyone and all aspects of an organization in continuous improvement through teamwork.'

The application of all the various quality management techniques and systems and the involvement and motivation of the people within an organization represent a significant management challenge requiring considerable resources. So why have so many organizations throughout the world identified quality development as the number one business priority?

The simple answer to this is **competition**. There is an old saying which states that in business there are only two problems – 'customers and competitors'. This message is well summarized in the often quoted statement attributed to Dr Edwards Deming, who said:

> In the future there will be two kinds of company – those who have implemented Total Quality and those who have gone out of business. You do not have to do this – survival is not compulsory.

The timescales for this message are certainly not the same across all sectors of industry and commerce but the principle remains the same. As markets, products and technology mature then the basis of competitive advantage is quality. In the long term, organizations compete on the basis of their product or service quality; TQM is therefore seen as an approach to gaining or sustaining a competitive lead.

Certainly there are many examples of major (often international) organizations who have been able to demonstrate that quality development and business development go hand in hand. The other viewpoint, however, is that TQM is a 'winners' methodology only applicable to mature world-class organizations. If TQM is seen as being one end of the spectrum of quality development, then certainly there is a requirement for such organizations to have reached a collective and individual state of maturity. TQM organizations need excellent systems, excellent techniques and methods and, above all, excellent people. The internal (in terms of cost base) and external (in terms of market share) benefits of reaching the advanced state of quality development which is characterized by TQM are enormous.

All organizations, however, can benefit by quality development and the competitive pressures for quality in less mature markets or business sectors are just as important. Many of the total quality companies have reached their state of development due to a combination of:

- the pressure (usually cost driven) to improve internal effectiveness;
- the pressure (usually customer driven) to improve external performance.

Industries such as motor car manufacture, telecommunications, consumer electronics, etc., have faced intense international competition for many years and as a result many of the most mature, total quality organizations are to be found in these sectors. The need for quality development in other sectors of industry is just as important, however, and in many cases, it is the early recognition by senior management of the need to prioritize quality development that is the critical factor in ensuring long-term business viability.

For most organizations in the 1990s the only certainty about competition in the future is that it will become more fierce. As the European Union develops, as markets in South East Asia mature and as American businesses strive to retain their share of international markets so the competition for quality will intensify. In industry, many companies are now explicitly seeking fewer suppliers and increasingly selecting their suppliers on their ability to produce quality products on time.

5.1.2 THE CHARACTERISTICS OF TOTAL QUALITY

So what does a total quality organization look like? Certainly companies that have adopted TQM are:

- improvement orientated;
- customer focused;
- quality driven.

But what do they actually do? What are the sort of activities which are taking place in TQM organizations that are absent from less developed ones? In simple terms these total quality organizations are using team-based quality management systems and techniques and they are involving their people in the process. The people in total quality organizations have two jobs:

- job number one is the work they do;
- job number two is finding ways of improving job number one.

This passion for improvement is part of the organization developing an improvement orientation (as described in Chapter 1). As the organization develops yet further and achieves a prevention orientation then a passion for the customer emerges and two rules are adopted.

- Rule number 1: the customer is always right.
- Rule number 2: if the customer is wrong, see rule number 1.

In other words, the organization always behaves in such a way as to promote excellence in customer service.

Internally, total quality organizations are characterized by their emphasis upon teamwork and a management style which is enabling rather than controlling. Quality improvement is seen as part of the process of running the business rather than as a separate activity. Decisions are made from the viewpoint 'does this enhance quality' rather than simply 'does this cost less'.

Externally, total quality organizations are generally able to charge more for their products (research shows that typically this premium pricing is around 6%) and enjoy a greater customer loyalty. Everyone in the organization is focused upon the customer, seeking to understand and satisfy the requirements and monitoring and improving performance.

5.1.3 QUALITY ASSESSMENT MODELS

One of the main problems with quality development is that there is no single, prescriptive approach. The tools, techniques, systems and methods described throughout this book have to be applied in an appropriate way to suit the business priorities and the organizational challenge. The reason that there is no single route to quality development is that there is no single starting point nor a single destination. Certain elements of total quality, such as quality system implementation through ISO 9000, have a very prescriptive approach (as described in Chapter 2) and a defined finishing point (accreditation). Similarly the stages in the implementation of SPC are well established (as outlined in Chapter 7) and the final outcome is easily recognized (capable processes). Total quality management is a mature integration of systems, techniques and behaviour and consequently no single assessment criterion is available or appropriate.

There is a need, however, to be able to validate an organization's claim to total quality and also a need for companies to measure their own comparative level of quality development. In response to these needs a number of international quality assessment mechanisms have been developed. The three most important of these are:

- the European Quality Award (primarily used in Europe);
- the Malcom Baldrige Award (primarily used in the United States);
- the Deming Award (primarily used in Japan).

The European Quality Award

The **European Quality Award (EQA)** is a quality assessment scheme developed and managed by the European Foundation for Quality Management (EFQM). The assessment criteria are divided into five basic areas as illustrated in Figure 5.1.

The EQA classifies the areas of assessment into either **enablers** or **results**. The enablers are the business activities which are employed to produce the results. This represents an important approach to self assessment by encouraging organizations to understand, appreciate and balance the efforts which are put in and the performances which result.

Each of the nine elements of the EQA model is assessed through a comprehensive self-appraisal questionnaire to measure an organization's quality development.

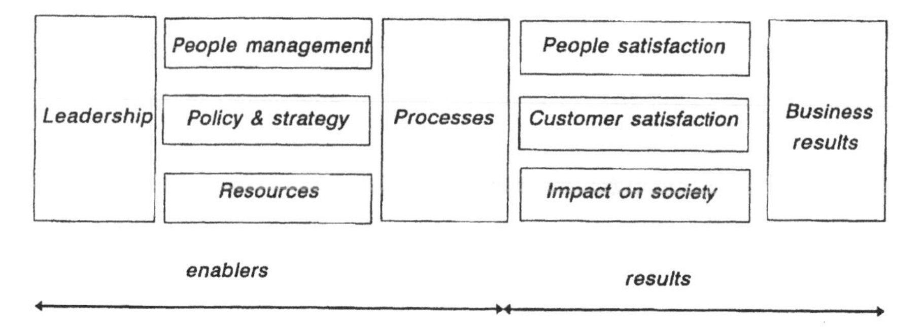

Figure 5.1 The European Quality Award assessment model.

The Malcom Baldrige Award

The **Malcolm Baldrige Award** (named after the US Secretary for Commerce) was established in 1987 to promote quality improvement and recognize achievements amongst American companies in the way the Deming Award had in Japan. The Baldrige Award assessment is based upon seven categories divided into leadership, system, measures and goals as shown in Figure 5.2.

In essence the Baldrige Award assesses quality development by evaluating the following.

- **Leadership** – not only in terms of the role of senior managers in the management of quality but also the way in which quality policy is communicated throughout the organization and the way in which the quality responsibilities to the community are effected.
- **Information and analysis** – the way in which information on quality performance is used to support the development of the organization, in

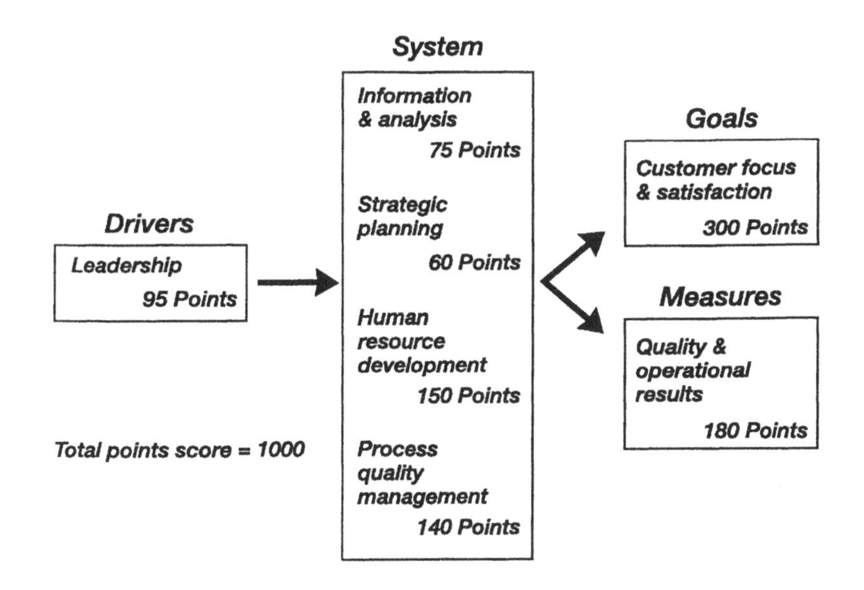

Figure 5.2 The Malcolm Baldrige Award assessment model.

particular competitive comparisons/benchmarking and the company-level use of data.

- **Strategic planning** – the extent to which quality improvement plans are incorporated into the business planning process and the use and development of quality plans to sustain quality development.
- **Human resource development and management** – the development, involvement, education, performance, recognition and satisfaction of employees.
- **Process quality management** – evaluation of the processes used within the company and the assessment of supplier organizations' processes to ensure ever-improving quality.
- **Quality and operational results** – assessment of the quality performance of the organization in operational terms and evaluation of how these results are used for comparison with competing companies.
- **Customer focus and satisfaction** – the most heavily weighted category which examines the relationships and commitment to the customers and evaluates customer satisfaction.

The United States Department of Commerce publishes guidelines for the evaluation criteria in each of the sections and each year two awards are made in each of the three categories of manufacturing, small businesses and service organizations. The Baldrige Award is seen as a broad measure of quality development.

The Deming Award

The **Deming Award** uses ten categories of quality activity as illustrated in Figure 5.3.

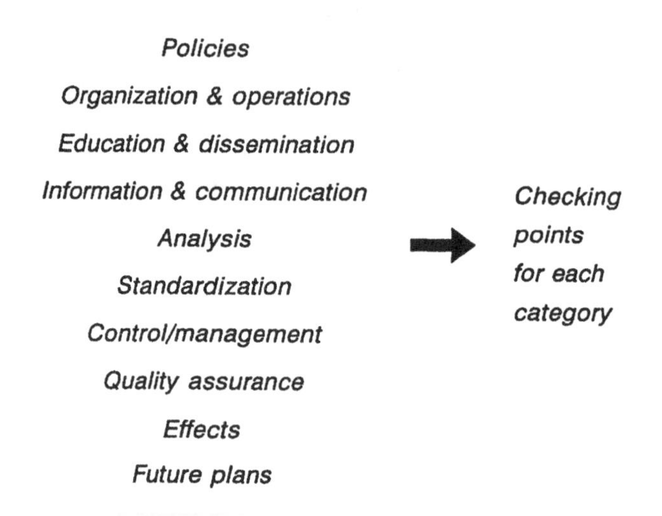

Figure 5.3 The Deming Award checklist.

- The policies are assessed by reviewing the existing management policy for quality and the methods used for establishing, communicating and implementing that policy.
- The organization is assessed in terms of the clarification of responsibilities, the interdepartmental coordination and the organization of the quality development activities such as steering committees and improvement groups.
- The education activities evaluated include the planning and management of employee education, particularly in statistical techniques and improvement methods, and also the training of supplier companies.
- The assessment of the collection and use of information within the organization is evaluated in terms of the quality of communications and analysis of statistical data. The evaluation of the analysis methods also examines the utilization of statistical methods and the pinpointing of quality improvements.
- The standardization category examines the establishment and application of both systems standards and technical standards.
- The assessment of control/management includes a review of management systems, control points, the application of control charts and the contribution to quality circles.

- Under the heading of quality assurance the product and service development activities, including preventative measures, are assessed together with customer satisfaction, process capabilities and the quality assurance system procedures.
- Finally the organization's future plans are assessed in terms of problem reduction and longer-term planning.

5.1.4 SUMMARY

- Quality development is fundamental to the long-term business success of all organizations.
- Total quality companies employ an integrated approach to systems, techniques and organizational behaviour.
- A number of international quality development assessment mechanisms have been developed including the European Quality Award, the Malcolm Baldrige Award and the Deming Award.

5.2 The basic elements of total quality management

5.2.1 THE MAIN COMPONENTS OF TOTAL QUALITY

Whilst there is no single blueprint for how total quality should be developed within an organization, there are certain fundamental components which need to be present. In terms of approach TQM requires:

- senior management leadership;
- improvements orientation;
- customer focus;
- company-wide involvement;
- Commitment to training and education;
- Ownership of the process;
- Emphasis on measurement and review;
- Teamwork.

In terms of the activities these can be categorized into seven main components as shown in Figure 5.4.

Organizations need to develop in each of the areas shown in Figure 5.4 in order to exhibit total quality although progress in each of these elements will not necessarily be equal as organizations prioritize their development activities.

One of the key components is therefore ownership of the quality development process. Planning for total quality requires people to understand the quality development process and to develop their own way forward on the basis of business priorities, customer needs and the prevailing management culture.

Figure 5.4 The main elements of total quality development.

5.2.2 THE DIMENSIONS OF QUALITY DEVELOPMENT

As outlined in Chapter 1, quality development is a three-dimensional process. Only if progress is made in all three aspects of development can total quality be used to describe the outcome. It is this problem of simultaneously progressing in three different dimensions that makes progress towards total quality so organizationally challenging. It is because each of the three dimensions supports the progress in the other two that this integrated approach is so important, as illustrated in Figure 5.5.

If progress is made only in developing the behaviour (people) then the quality development will be naive and will not provide sustained improvement in business performance. Conversely if the developments only concern the systems and the techniques then the development will be brittle and when the business comes under pressure (say, from competitor performance) then the attitudes may not support the organizational changes required. The skills required to simultaneously progress in each of these three dimensions are unlikely to be present in a single individual and hence company-wide involvement and a team-based approach are essential.

The management challenge with total quality is the simultaneous development of people, systems and techniques. The difficulty of this task requires sustained management commitment over a period extending beyond the next balance sheet. The long-term business benefits from the synergy of total quality are enormous in terms of operating efficiencies, customer satisfaction and employee motivation.

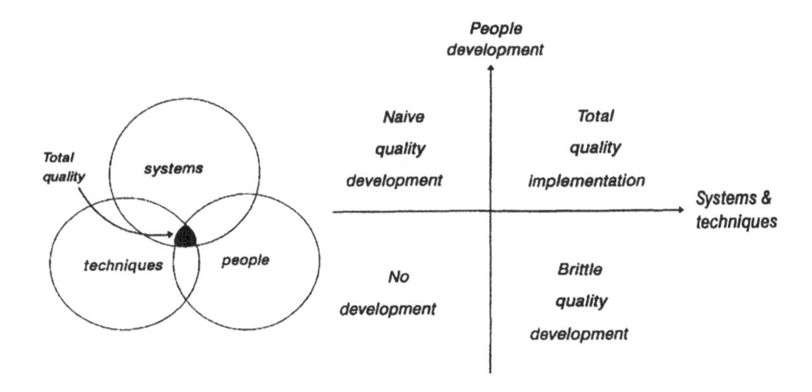

Figure 5.5 The dimensions of quality development.

5.2.3 SUPPLIER DEVELOPMENT

The development towards total quality management involves the improvement of both internal and external business processes as illustrated in Figure 5.6.

The development of a customer orientation is therefore a critical aspect of total quality as described below in section 5.4. Also important, however, is the development of suppliers particularly when considering that typically:

- over 50% of the manufacturing costs are due to raw materials and bought-in items;

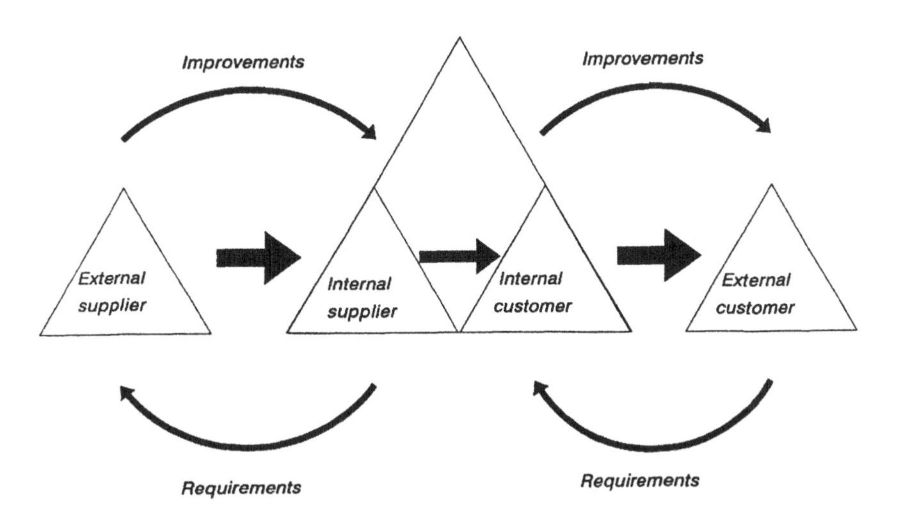

Figure 5.6 The internal and external customer–supplier chain.

- over 60% of customer complaints on complex products (motor cars, consumer electronics) are due to defects in subcontracted materials or components.

The development of suppliers is therefore an important element of total quality and a number of techniques have been developed to assist in this process, including:

- supplier rationalization;
- supplier partnerships;
- supplier assessment and vendor rating;
- supplier education;
- supplier awards schemes.

Supplier rationalization

Supplier rationalization involves the adoption of fewer, better suppliers. Many organizations as part of their quality development policy have reduced their supplier base typically by 60–80%. Instead of having four or five suppliers competing to supply on price, it is much better to have only one or two suppliers who optimize the overall cost of ownership of the raw material. Concern is sometimes expressed that having only one or two suppliers of a critical raw material of a component may leave the organization vulnerable both commercially and logistically. In most cases this is not the case as dealing with fewer suppliers enables much closer relationships to be developed and allows time for the effective planning of material supply (which generally has a greater impact upon the reliability of supply than does additional numbers of suppliers).

Supplier partnerships

Supplier partnerships (sometimes referred to as **co-makership**) requires a much closer and earlier involvement of suppliers who should be viewed as an extension of one's own business rather than an outside organization with whom to 'do battle'. The foundation of partnership is a clear requirement as it is estimated that around 50% of the queries/rejections on incoming materials are due to unclear requirements. In order to develop, organizations need to put proportionately more of their effort into clearly defining and agreeing requirements with suppliers rather than the traditional adversarial relationship of inspection–rejection. Many quality developed organizations form improvement teams with their suppliers to clarify requirements, identify improvements and to implement corrective actions. As organizations mature to a prevention orientation, then suppliers are increasingly involved in the design process to ensure the requirements of their materials are clearly understood and that they are capable of providing reliable inputs to new products or services.

Supplier assessment and vendor rating systems

Supplier assessment and vendor rating systems are generally used as part of the development of a systems orientation and typical of the assessment approach to supplier development is ISO 9000 as described in Chapter 2. Vendor rating schemes normally provide some form of scoring mechanism based upon the supplier's performance in terms of product quality, delivery, service and price. Typical vendor rating schemes are shown in Figure 5.7.

	Quality	*Delivery*	*Service*	*Price*
Mark out of	100	100	100	100
Weighting factor	0.7	0.15	0.05	0.1

Supplier categories	
	A = 80 and above
	B = 50 to 80
	C = Below 50

Figure 5.7 Typical vendor rating scheme.

Supplier education

Supplier education can take the form of seminars and workshops in which suppliers are made aware of the methods, techniques, requirements and expectations of the purchaser. Typical of this approach are supplier training programmes in the application of statistical process control (described in Chapter 7) whereby supply companies are encouraged to use statistical methods and to demonstrate process capability.

Supplier awards

Supplier awards are used to reward suppliers who have achieved certain quality, service and delivery goals during the year. This recognition can be used by suppliers to promote their products or services with other customers.

5.2.4 SUMMARY

- The main components of total quality include leadership, continuous improvement, customer orientation and company-wide employee involvement.
- The process of quality development requires simultaneous progress in terms of systems, people and techniques.
- External improvements including supplier development are integral to the implementation of total quality throughout the supply chain.

5.3 Alternative approaches to total quality management

5.3.1 TOTAL QUALITY MANAGEMENT IMPLEMENTATION METHODOLOGIES

As stated above, there is no single route to total quality because there is no single starting point. As yet, there is no internationally recognized 'standard' approach to quality development although it is possible to identify key implementation factors which distinguish one approach to TQM from another. Research has shown that the three key factors are:

- organization;
- approach;
- motivation.

The **organization** of the total quality implementation programme can take a number of different forms. Among the critical organizational decisions are:

- How should the TQM process be steered?
- What type of improvement teams should be employed?
- What should be the mechanisms for creating culture change?

The **approach** adopted to the implementation of total quality focuses upon the main vehicles used in changing the way quality is managed within a company. Among the approaches which can be adopted are:

- techniques led – focusing upon the application of tools such as problem-solving, statistical process control or quality function deployment;
- 'guru' led – whereby 'established' programmes generally promoted by consultants who may have observed a number of quality development programmes and identified generic implementation approaches;
- vision led – in which the mission for the organization (for example, customer satisfaction) is enthusiastically communicated to all employees and becomes a focus for the quality developments within the company.

The **motivation** for the total quality implementation can be provided from a number of (not necessarily mutually exclusive) sources including:

- leader driven – normally from the chief executive;
- champion driven – normally the quality manager;
- quality council driven – led by a group of individuals;
- corporately driven – from external corporate level objectives.

The final major influence on the implementation of total quality is the **timescale** adopted by the organization. This timing element is in fact perhaps the most critical dimension of implementing total quality. The two most commonly observed pitfalls in terms of implementation timescales are as follows.

- Organizations employ a cautious, gradual approach whereby the understanding of the principles of total quality are allowed to slowly permeate down through the organization (usually in the form of training or awareness seminars) and the cultural changes are allowed to occur over some 'natural' timescale. Typically such an approach begins to peter out after about two years as the 'critical mass' needed for quality development never materializes and therefore the expected benefits are not realized and senior management enthusiasm wanes.
- Organizations employ a vigorous, breakneck approach whereby bold improvement objectives are set and employees are required to immediately become involved in the quality development process. Such an approach is often driven by some form of major organizational crisis (for example, market share or financial performance disasters) and total quality is seen as some form of panacea. With this type of implementation timescale, typically the quality development process 'stalls' and the discontinuity between the traditional approach to managing quality and the newly implemented total quality ideology come into conflict.

Organizations have an innate inertia and therefore the process of driving the change towards total quality should be applied at a rate which is fast enough to prevent the development grinding to a halt, but not too fast to create an insurmountable 'back-pressure'.

5.3.2 ORGANIZATION AND PLANNING FOR TOTAL QUALITY MANAGEMENT

Quality development is a management responsibility and as such, therefore, requires to be planned and organized. At the highest level, the planning for total quality should define the organization's:

- mission;
- values;
- measures.

The organization's **mission** should be a clear statement which can be understood by all employees and can provide a focus for the quality developments within the company. Mission statements come in a variety of forms, from simple competitive war-cries such as 'To be the world's leading supplier of ...' or 'To be the best of the rest outside Japan' to rather more customer-led statements such as 'To be recognized as the most responsive, customer-orientated supplier of ...' or 'To excel in quality service and hygiene'. Whilst such mission statements are sometimes viewed cynically as 'corporate hype' if properly formulated they do represent an important management tool for aligning the culture of the organization with the business objectives.

The **values** of the organization should also be clearly stated and express the principles upon which the mission will be achieved. Typical values statements include the following.

- Value leadership, integrity and teamwork.
- Our people will be provided with the tools, timing and time to provide quality work.
- Value the contribution of individuals.
- Nobody's job will be a victim of improvement.

Finally, the *performance* measures which are defined become the key indicators of quality development. On the basis of 'What gets measured gets done' the planning of performance measures provide an important linkage between the quality development efforts and the business objectives.

In the early stages of quality development, establishing a formal organizational structure to deploy objectives and to monitor and review progress is also an important prerequisite for success. Typically organizations implementing total quality form a company-wide quality steering group (or council or committee, the terminology varies) comprising the chief executive and senior managers. The main responsibilities of this senior quality steering group are to define and prioritize the quality improvement objectives and to maintain the correct rate of implementation (discussed above in section 5.3.1). For organizations with more than one operating location then a site quality council is usually organized with responsibilities for deploying the improvement objectives down to improvement projects and to manage and coordinate the activities of the quality improvement teams. The site quality council ensures that the rate of total quality implementation is properly resourced and also ensures that improvement team members receive recognition for their contributions and that recommended improvements are implemented. Finally improvement teams are organized with leaders, facilitators and members as described in Chapter 4. A typical structure of the organization required for the implementation of total quality is shown in Figure 5.8.

As the quality development process matures then the quality improvement organizational structure becomes an integral part of the management and operational processes within the company.

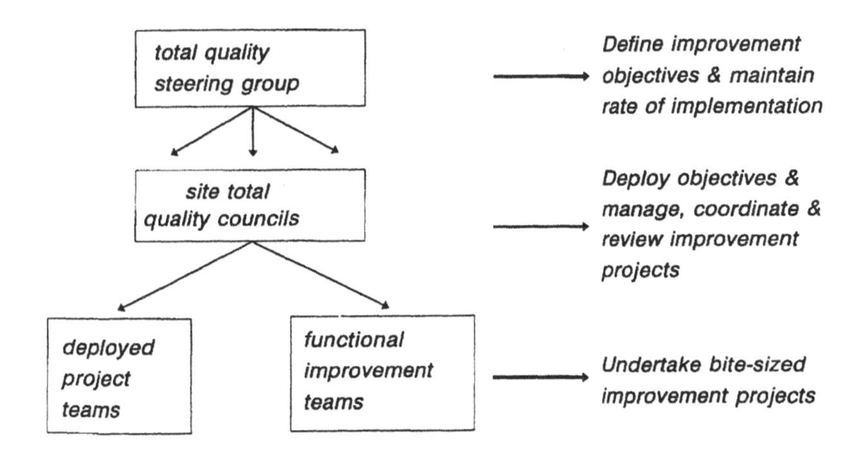

Figure 5.8 Organization for implementing total quality.

5.3.3 TOTAL QUALITY MANAGEMENT IMPLEMENTATION PLAN

The basic framework for quality development outlined in Chapter 1 involves establishing a systems orientation, an improvement orientation and finally prevention orientation. The implementation plan to be adopted by any given organization needs to be self-generated and is a responsibility of senior management. Whilst there is no single blueprint or 'route map' there is a certain set of generic stages as illustrated in Figure 5.9.

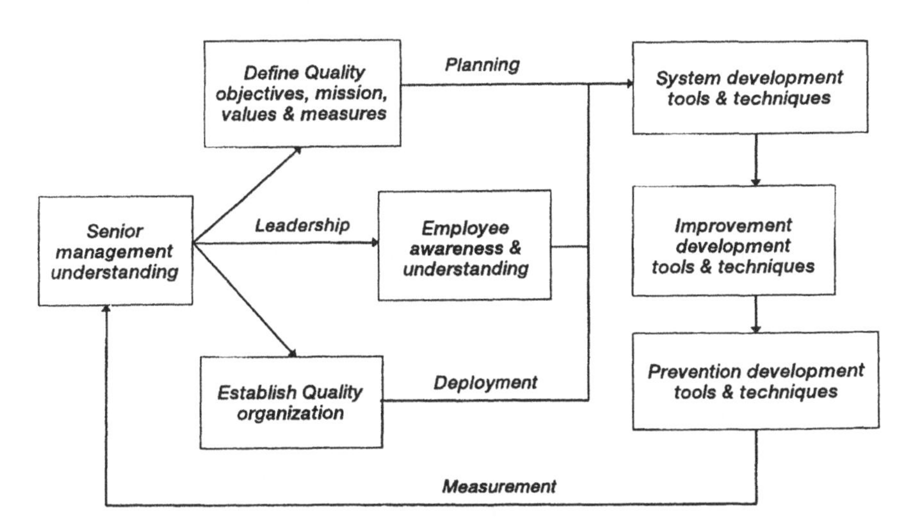

Figure 5.9 Total quality management implementation plan.

The emphasis, in terms of the approaches and techniques to be used and the timescales employed, will differ from organization to organization depending upon the type of organization, the nature of the business and the development objectives. The key issue is for the organization to understand the quality development process, to understand the role of the various techniques, to have a clear view of the quality objectives and to own the total quality implementation plan.

5.3.4 SUMMARY

- The implementation of total quality requires appropriate organization, techniques and motivation, and the rate of implementation needs to be carefully managed.
- Initially total quality implementation requires a senior level steering group and a local quality council to plan and coordinate the efforts of the improvement teams.
- Implementation requires planning and ownership of the quality development process.

5.4 Developing a customer orientation

5.4.1 CUSTOMERS AND TOTAL QUALITY MANAGEMENT

Many organizations are capable and indeed successful in developing quality systems and achieving the first stage of quality development. Of these organizations, research shows that the majority do not progress beyond this stage of development and remain with quality assurance as the primary approach to managing quality. Those organizations which progress towards an improvement culture generally focus upon the internal business processes in order to reduce lead times, reduce waste and inefficiency and to increase product or service performance. Again a relatively small proportion of companies progress beyond this internal improvement focus and become prevention orientated by designing robust and reliable products and processes. Perhaps the most critical factor in providing the impetus to achieving this final stage of total quality development is customer orientation. Focusing upon the customer and customer service is a prerequisite to developing and sustaining a total quality culture.

The importance of meeting customer requirements is of course a fundamental element of developing quality systems. Similarly many quality improvements benefit the customer directly and an organization whose products and services are continuously improving will clearly enjoy the market benefits associated with improving customer satisfaction. Beyond customer requirements and customer satisfaction, however, is a concept of customer delight whereby the organization develops a passion for providing quality products and services to and relationships with its customer. The aim of this final stage of quality development is to create customer loyalty. This loyalty is generated through orientating the whole organization towards serving the customer. The business importance of customer loyalty has been identified by considerable research which has shown in general:

- over 60% of an organization's future revenues will come from existing customers;
- a 2% increase in customer retention has an equivalent impact upon profitability as a 10% reduction in operating costs;
- up to 96% of unhappy customers do not in fact complain, but they are three times more likely to communicate a bad experience to other customers than a good one;
- if a customer complains and the organization responds effectively to the product or service failure, then the loyalty of the customer can actually increase.

Developing this customer orientation and customer loyalty is key to maintaining organizational progress beyond the improvement stage. Many total quality implementation programmes begin to falter after a couple of years of undertaking team-based improvement projects and the key to further development is to address the cross-functional issues which improve customer service.

Developing customer loyalty is, however, a major organizational challenge and requires significant cultural change. Meeting basic customer requirements can be readily achieved through the mechanistic application of quality systems. An improvement orientation employing customer service improvement projects can develop customer satisfaction but the achievement of customer loyalty requires the creation of a culture whereby commitment to the customer is exhibited by all employees and the importance of the relationship with the customer is recognized.

5.4.2 INTERNAL AND EXTERNAL CUSTOMER FOCUS

Excellence in (external) customer service can only be achieved if the internal customer–supplier chain operates effectively. The problem in many organizations is that many individuals and functions within the business are quite remote from the external customer. Encouraging individuals to have a 'passion' for the customer may be irrelevant in a functionally arranged organization where employees have little or no contact with the customer. The external improvements which stem from more clearly defining specifications for suppliers or from better understanding the customers' requirements can also be translated to the internal business processes.

The concept of internal customers and suppliers is an important quality development tool. By emphasizing better internal customer service, everybody in the organization can become part of the culture change of increased customer focus. Two techniques are particularly useful in developing the internal customer–supplier chain:

- internal customer–supplier requirements definition;
- department purpose analysis

The definition of internal customer–supplier requirements is a useful exercise for individuals to list who they believe their key internal customers and suppliers are and what are their prioritized requirements. These requirement definitions can then be compared with the corresponding requirements defined by the internal customers and the perceptions of the internal suppliers. By performing a gap analysis on any major differences between the internal customer and the requirements perceived by the internal supplier provides an important improvement opportunity to align the business processes more effectively and to improve the service to the external customer. The process of internal requirements definition is illustrated in Figure 5.10.

Departmental purpose analysis (DPA) is a technique developed within IBM which encourages departments or work groups to formally document the 'purpose' of the work carried out. The DPA is carried out by the departmental group led by the manager, who brainstorm the departmental tasks, prioritize the main tasks and identify the main internal customers and suppliers of each activity. These activities and departmental priorities are then reviewed with each customer or supplier in turn to evaluate the extent to which needs are being met and measures for conformance established. The DPA process is intended to generate improvement opportunities and also

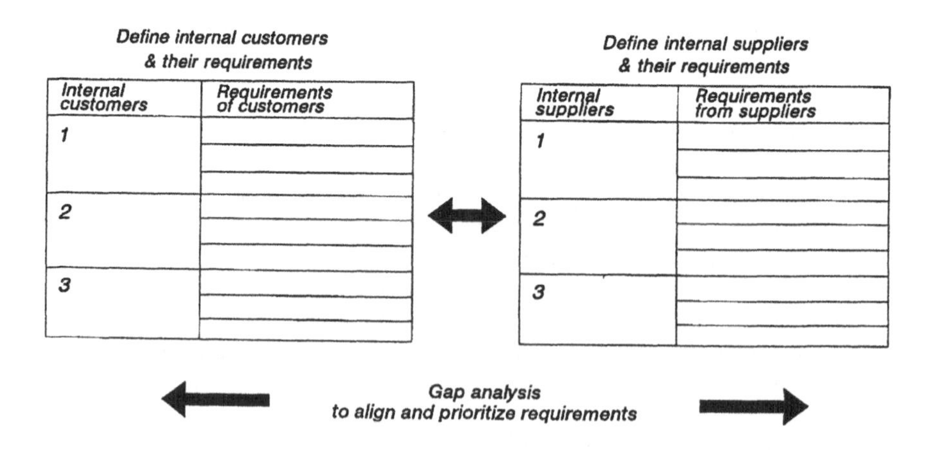

Figure 5.10 Internal customer–supplier requirements definition.

generate a common, validated view of the departmental purpose. The DPA documentation shown in Figure 5.11 should be regularly revisited to alert the organization to changes in emphasis or priorities.

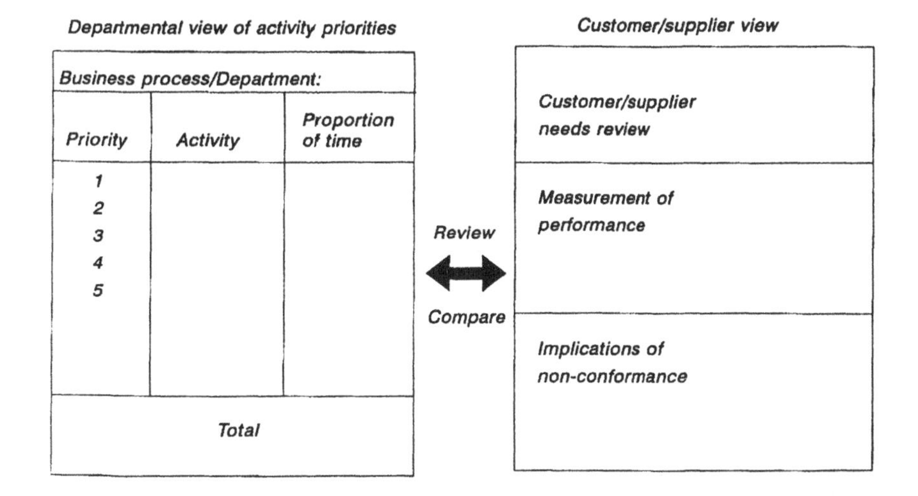

Figure 5.11 Departmental purpose analysis.

5.4.3 DEVELOPING CUSTOMER SATISFACTION

The development of customer satisfaction can be mapped onto the quality development process described in Chapter 1 as shown in Figure 5.12.

The cultural change needed to promote customer satisfaction commences therefore with the fairly mechanistic systems activities of defining requirements and incorporating these refinements into the business procedures. The second stage involves the measurement of the customer service performance and the implementation of team-based improvement activities to enable the organization to exceed the customer's basic requirements. During the final stage the emphasis is upon behaviour and the relationships developed with customers which promote increased loyalty.

A customer can recognize a total quality organization almost immediately from the manner in which the customer is dealt with. Total quality organizations make the customer feel 'special' and deal with the customer as a matter of priority rather than as a secondary issue, less important than the undertaking of other business activities. Nothing is more infuriating for a customer than to be kept waiting by an organization who instead of serving the customer continues with an alternative, internal activity. For most organizations the importance of the customer lies in the fact that the only source of income for the business is customers.

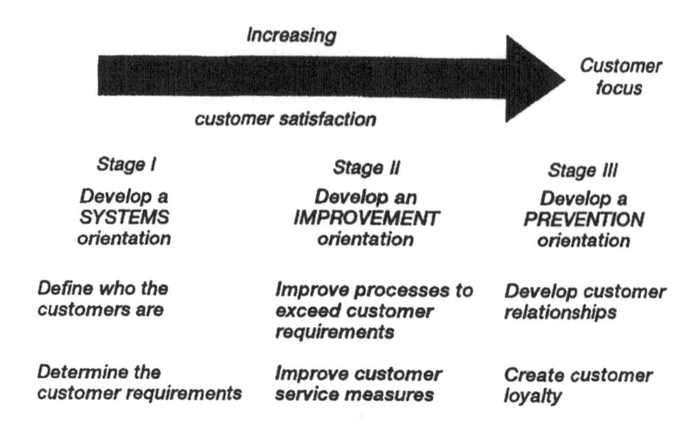

Figure 5.12 Developing customer satisfaction.

5.4.4 SUMMARY

- Progress towards total quality requires an organization to develop a customer-orientated culture.
- External improvements in customer service require internal developments using techniques such as internal customer–supplier requirements definition or departmental purpose analysis.
- The stages in developing customer satisfaction involve establishing and meeting customer requirements, improving the internal processes to exceed customer expectations and finally building customer loyalty.

5.5 Business process re-engineering

5.5.1 THE ROLE OF BUSINESS PROCESS RE-ENGINEERING IN TOTAL QUALITY MANAGEMENT

Fundamental to the development of a continuous improvement culture is the involvement of all employees in incremental improvement projects. This philosophy of continuous improvement through a large number of small improvements is encapsulated in the Japanese concept of 'kaizen' as described in Chapter 8. This approach to quality development does, however, assume a certain timescale and rate of total quality implementation. For many organizations, business pressures require rather more significant quality developments and rather quicker timescales. The approach to fundamentally redesigning the business processes rather than simply improving the operation of existing processes is termed **business process re-engineering (BPR)**.

A commonly used definition of BPR is:

> The **fundamental** rethinking and **radical** redesign of business processes to achieve **dramatic** improvements in critical contemporary measures of performance such as cost, quality or service.

The essence of BPR is therefore changing the organization from a functional arrangement to a process basis in which the work groups carry out multi-dimensional tasks associated with a business process rather than simple tasks associated with a business function. So, for example, rather than attempt to incrementally improve each stage of the product design activity the process could be re-engineered to concurrently engineer the new products. BPR typically involves organizational change where employees in their process rather than functional roles are empowered to take more responsibility and to carry out more complex tasks. The performance measures associated with BPR are therefore more concerned with business results (for example, customer service levels) than with business activities (such as the number of orders processed).

Kaizen and BPR are not necessarily mutually exclusive total quality activities but relate to different rates of change and to different stages of development.

5.5.2 IDENTIFYING BUSINESS PROCESS RE-ENGINEERING OPPORTUNITIES

One of the key initial stages of BPR is actually identifying what constitutes a business process. Having studied the organization and identified the business processes the next key question is which processes require to be re-engineered?

The identification of business processes can be more difficult than would perhaps be anticipated. Organizations which are functionally arranged in order to create specialized activities very often mask the underlying business processes. In simple terms business processes can be classified as either:

- **delivery processes**, which directly interact with the customer (for example, order processing, product delivery or product maintenance); or
- **support processes**, which sustain the delivery processes (for example, materials acquisition, financial planning or recruitment).

The critical business processes are identifiable by having:

- measurable inputs and outputs;
- value adding properties;
- activities which are repeated.

One of the key facets of business processes is information flow. If a data flow diagram is constructed for an organization the key processes emerge as being concentrations of data processing. These critical business processes require proper planning in terms of process definition and objectives and effective control in terms of process measures and ownership.

To identify which processes require re-engineering a number of useful criteria can be applied:

- processes which are disjointed and hence contain excessive information exchange, excessive inventories or buffer stocks or excessive amounts of checking and rework;
- processes which add significant value to the customer and impact upon the quality development process;
- processes whose re-engineering is likely to succeed.

In the re-engineering of processes generally the scope and responsibilities of the people involved increase. To facilitate the increased responsibilities of the individuals involved in the re-engineered process normally some form of restructured information processing is required. Typically BPR involves the application of information technology to more effectively manage the access and use of data within the boundaries of the re-engineered process. Instead of the traditional management approach of simplifying the tasks and increasing the complexity of the interfaces, BPR involves creating more complex, holistic processes with simpler interchanges.

5.5.3 IMPLEMENTING BUSINESS PROCESS RE-ENGINEERING

As with many of the advanced quality development techniques, the implementation of BPR is primarily a team-based activity. Implementing BPR involves a two-stage process, firstly identifying the top-down re-engineering opportunities and secondly the identification of process improvements.

The initial stage, normally undertaken by a process re-engineering team in conjunction with senior managers, identifies the key business processes and the improvement objectives. One of the most quoted examples of effective process re-engineering occurred at the Ford Motor Company which, in investigating improvement opportunities in material procurement, evaluated the equivalent processes at Mazda in Japan. Instead of attempting to incrementally improve the system for paying suppliers' accounts, Ford re-engineered the process by authorizing payment at the point of receipt rather than at the point of invoice as shown in Figure 5.13.

Having identified the improvement objectives and measured the 'gap' in terms of performance compared to customer requirements (or competitor's performance) the second stage is to establish a process improvement team. The improvement team would normally:

- analyse the existing process in terms of activities and performance;

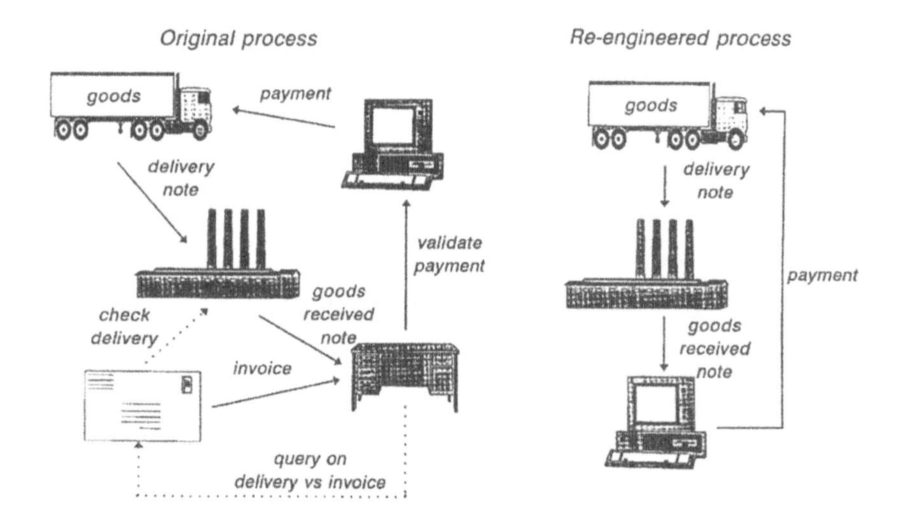

Figure 5.13 Process re-engineering of accounts payable.

- benchmark the process against best business practices;
- implement the re-engineered process.

The implementation of re-engineered processes can, however, involve significant organizational and technological change. The management of such change can often be a major organizational challenge and can take a considerable length of time. In the words of Robert Kennedy. 'Progress is a nice word but change is its motivator and change has its enemies.'

5.5.4 SUMMARY

- Business process re-engineering can provide significant performance improvement through a change from functional to process management.
- Business processes can be classified as either delivery of support processes and can be re-engineered through the simplification of the interfaces, often through the creation of more complex empowered activities.
- The implementation of BPR is team-based and involves an identification stage and an improvement stage.

6 Acceptance sampling

6.1 Sampling or 100% inspection?

6.1.1 THE ECONOMICS OF ACCEPTANCE SAMPLING

The subject of inspection and its frequency generates a certain amount of disagreement between quality management specialists. Typical viewpoints from this debate are:

- 'If we are serious about quality, then we should inspect every item.'
- 'The need for subsequent inspection surely represents a failure of our system of assuring quality.'
- 'Inspection is the domain of managing quality through quality control and is inconsistent with the philosophy of total quality management.'

The assumption inherent in the first statement is that inspecting every item (100% inspection) is 100% accurate. Even under circumstances where the inspector is adequately trained to carry out the task required, mistakes can arise for a variety of reasons associated with the occurrence of human error. The reader of this book, having reached Chapter 6, could reasonably be expected to carry out a task requiring simple reading and counting skills. Examine the sentence in italics in Figure 6.1. and, allowing 10 seconds, count the number of times the letter 'f' is used. (Somewhat less than 10% of undergraduates can regularly carry out this 100% inspection 100% accurately.)

Allowing yourself 10 seconds, count the number of times the letter 'F' is used in the following sentence:

> Finished files are the result of years of scientific study combined with the experience of many years.

Correct answer = 6

Figure 6.1 100% inspection task.

On a more serious note, the limitations of 100% inspection are:

- not necessarily 100% accurate (typically 85% accuracy is achieved);
- not applicable where testing is destructive;
- may actually reduce the quality level where the testing is intrusive or disruptive;
- a reduced pressure to get things 'right first time';
- the cost may be prohibitive.

In evaluating the economics of adopting sampling rather than 100% inspection, consideration needs to be given for the balance between the cost of inspection and the cost of failure. The second of these two elements, the cost of failure, is normally a function of both the implications to the customer of the product or service failing and also the likelihood of failure occurring. In quality cost terms (Chapter 3) this is the relationship between the cost of conformance and the cost of non-conformance. The risks involved in sampling rather than 100% inspection are normally described as:

- **producer's risk (Type I error)** – the probability associated with concluding from the evidence of a sample that the batch is deficient when the actual batch quality is acceptable;
- **consumer's risk (Type II error)** – the probability that the evidence of the quality of the sample will lead to concluding that a batch is acceptable when in fact the overall quality level is deficient.

6.1.2 THE ROLE OF ACCEPTANCE SAMPLING IN QUALITY DEVELOPMENT

The questions raised in section 6.1.1 on the validity of using acceptance sampling are important to evaluate in terms of the role of sampling in quality development. The need to verify conformance to customer requirements arises at all stages of quality development.

Whilst the notion of 'acceptable quality level' is inherent to most of the techniques of acceptance sampling, this terminology is often criticized in that it implies a level of defectiveness being acceptable to impose upon the customer. However, if quality management is viewed as a developmental process, then quality levels (in terms of proportion defective) accepted as part of acceptance sampling can be continuously improved (reduced) and used as a measure of development. The contribution of acceptance sampling to each of the stages of quality development is illustrated in Figure 6.2.

6.1.3 THE APPLICATION OF SINGLE ACCEPTANCE SAMPLING

The basic pattern of acceptance sampling is as follows:

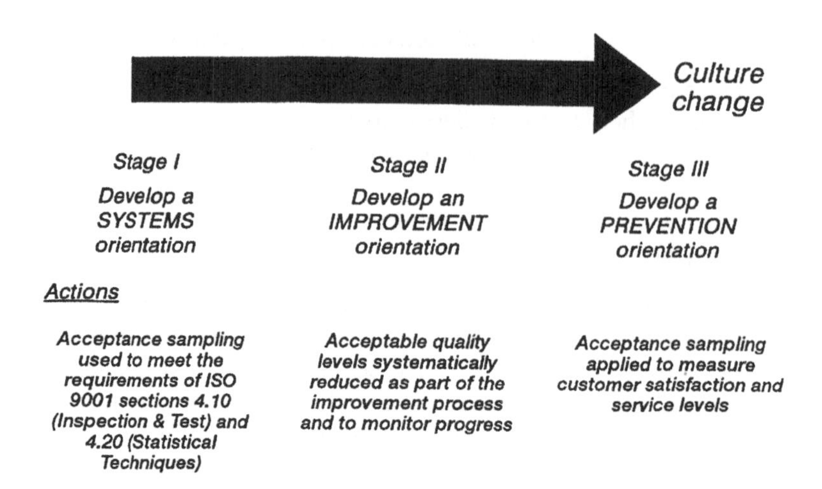

Figure 6.2 The role of acceptance sampling in the quality development process.

- A batch (or group) of products of size N is presented for inspection/evaluation.
- A small sample of size n is randomly selected from the batch and is subjected to the inspection/evaluation procedure.
- Providing the number of defects found in the sample does not exceed a specified limit of size c then the whole batch is accepted. If, however, the number of defects exceeds the value c then the batch is rejected (normally for 100% inspection by the original process owner).

The basic rules as to where to apply acceptance sampling are:

- after changes in responsibility;
- before high added-value processes;
- after processes which have a high or variable defect rate;
- where required by contract.

Having decided where in the business process chain acceptance sampling should be applied, the next consideration is normally what type of conformance information is required. The acceptance criteria can be either:

- **attribute data** – whereby the product or service is inspected/evaluated on a pass or fail basis (conforming or non-conforming);
- **variables data** – whereby the measured value of the inspected parameter is recorded and used as the basis of the inspection/evaluation.

The essence of acceptance sampling is a reduction in the amount of inspection or verification through the application of statistical inferencing based upon the evidence of a small sample. The quality levels against which the inferences are drawn are as follows:

- **Acceptable quality level (AQL)** is the defect level present in the complete batch that is the maximum level acceptable to the customer/marketplace as a process average.
- **Limiting quality level (LQ)** (or lot tolerance percent defective – LTPD) is the defect level in the complete batch that is deemed to be unacceptable to the customer/marketplace as a process average.

At first sight these quality levels may simply appear to be different ways of expressing the same concept, the point at which the quality level in the batch is either acceptable or not to the customer. In most practical instances, however, the point at which the product is rejected by the customer and the normally expected level of quality are at separate defect levels and the quality level between the two levels is described as indifferent.

In terms of the operation of acceptance sampling, the samples can be either:

- **single** – where one sample is taken from the batch and evaluated;
- **multiple** – where two or more separate samples are taken sequentially;
- **sequential** – where the sample is made up of individual elements selected over the life history of the batch.

The most common form of acceptance sampling is the selection of a **single** sample using **attribute** data.

6.1.4 SUMMARY

- Acceptance sampling is necessary where the cost of 100% inspection is unacceptable.
- Acceptance sampling can be used throughout the quality development process, in the establishing of quality systems, in the measurement of improvement and in the evaluation of customer satisfaction.
- Acceptance sampling is applied through the selection of the type of data and the appropriate quality levels.

6.2 The basic elements of acceptance sampling

6.2.1 OPERATING CHARACTERISTICS

The relationship between the defect level present in the batch and the probability that the batch will be accepted on the basis of the quality of a specific sample (size n and acceptance number c) is termed the operating characteristic (OC) curve.

The probability of accepting a batch on the basis of the quality of the sample can be determined providing the batch size is large in comparison to the sample size using the binomial distribution. If PD is the percentage defect present in the batch and P(r) is the probability of finding r defects in the sample then from the binomial distribution:

$$P(r) = \frac{n!}{r!(n-r)!}(\frac{PD}{100})^r(1-\frac{PD}{100})^{(n-r)}$$

For example, consider a batch where: PD = 10%, sample size, $n = 10$. What is the probability of accepting 1 or less defects?

$$\text{Probability of 0 defects, } P(0) = \frac{10!}{1(10)!}(\frac{10}{100})^0(1-\frac{10}{100})^{10} = 0.35$$

$$\text{Probability of 1 defect, } P(1) = \frac{10!}{1(9)!}(\frac{10}{100})^1(1-\frac{10}{100})^9 = 0.39$$

$$\text{hence } P(\leq 1) = P(0) + P(1) = 0.35 + 0.39 = 0.74$$

If the percentage defect in the batch is small then the Poisson distribution can be used as an approximation. This has the advantage in that the calculation of the probability can be simplified and these simplifications are published in chart form such as the modified Thorndike chart shown in Figure 6.3. This chart is named after its originator who continued the developments on acceptance sampling tables begun at the Bell Telephone Laboratories in the USA by Dodge and Romig.

For a specific acceptance sample in which the sample size, n, and the acceptance number, c, is defined, then the modified Thorndike chart can be used to determine the probability of accepting the batch for a range of values of percentage defective. Plotting this relationship generates the typical OC curve shown in Figure 6.4.

Also shown in Figure 6.4 is the relationship between the acceptable quality level (AQL) and the producer's risk (α) and between the limiting quality level (LQ) and the consumer's risk (β). By comparing the OC curve with the 'perfect' curve which would be generated through the application of 100% accurate, 100% inspection two areas can be identified:

- **area I** – which represents the cumulative probability of rejecting good batches;
- **area II** – which represents the cumulative probability of accepting bad batches.

6.2.2 AVERAGE TOTAL INSPECTION

Inherent in the probabilities of inferencing illustrated by the OC curves is the problem of the probability of acceptance being less than 1, even when the process is producing goods or services which are at or better than the acceptable quality level.

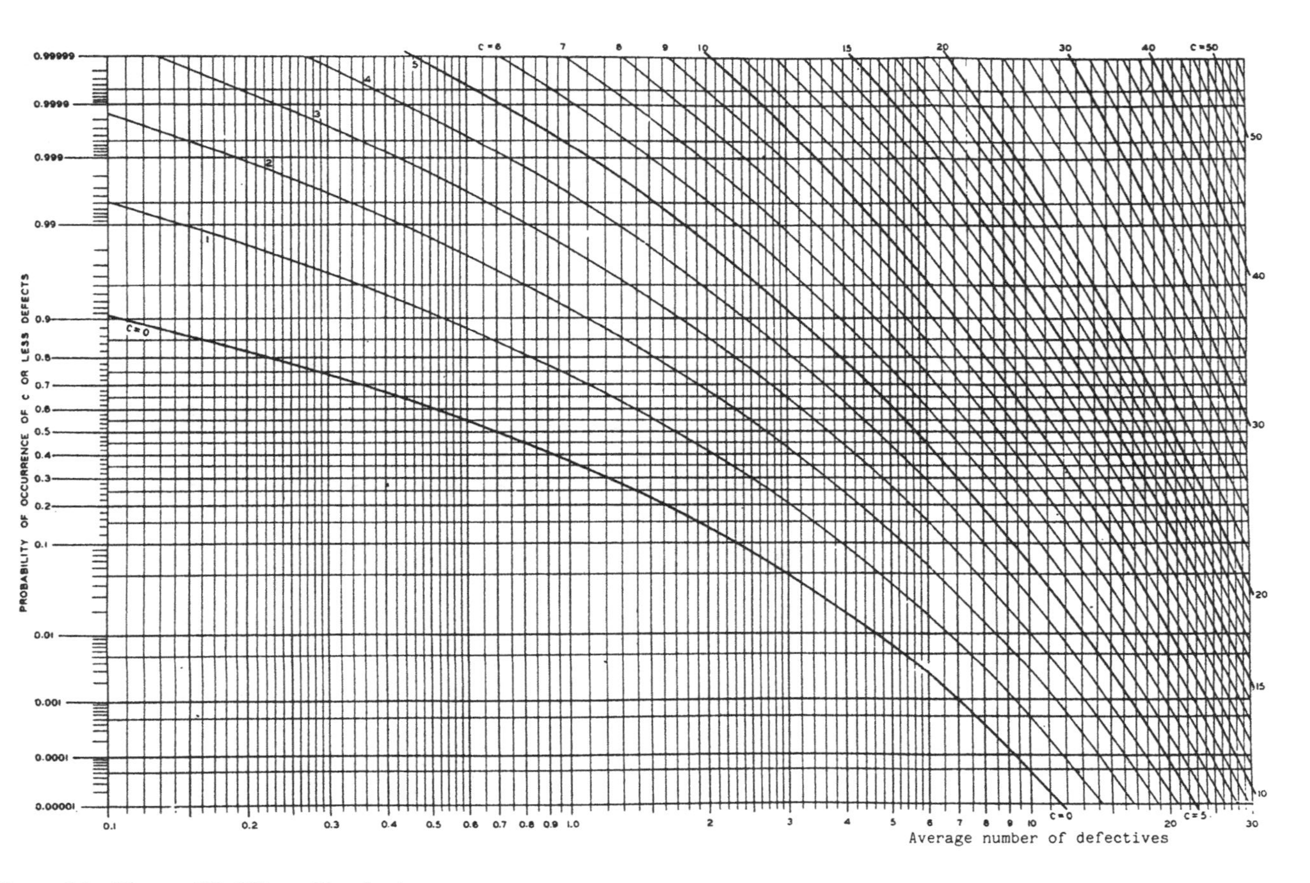

Figure 6.3 The modified Thorndike chart.

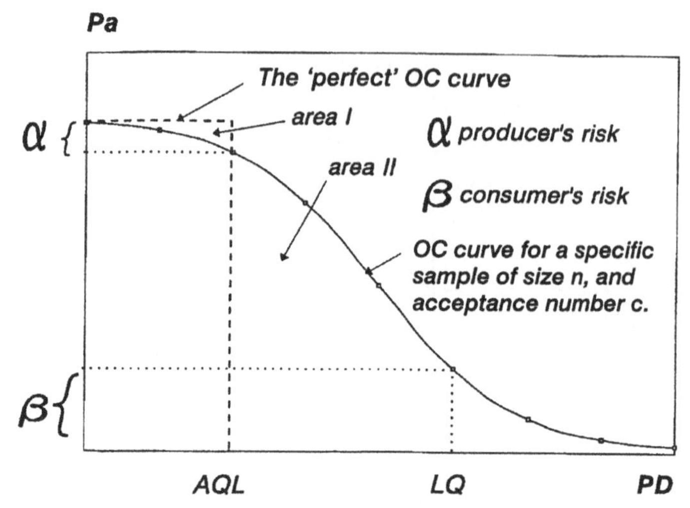

Figure 6.4 Typical operating characteristic curve.

Due to the probability of acceptance being less than 1, the total quantity of items which will need to be inspected over a number of batches is not simply the sum of the individual samples. Consider a sampling plan for which the probability of accepting the batch when the percentage defect at the AQL is 90%. Hence the probability of rejecting 'acceptable' batches at this quality level is 10%. Therefore the average number of items which would need to be inspected over 10 batches of size N would be:

$$(10 \times \text{sample size, } n) + (1 \text{ rejected batch of size } \mathrm{N} - n)$$

Despite this risk, the sampling plan still represents a significant reduction in the total number of items requiring inspection when compared to 100% inspection.

In general the average total inspection (ATI) is equal to:

$$\begin{aligned} \text{ATI} &= (\text{Sample size}) + (\text{Average number of items in} \\ &\quad \text{remainder of batches rejected due to producer's risk}) \\ &= n + (\mathrm{N} - n)\,\alpha \end{aligned}$$

The average total inspection should be used when calculating the total appraisal cost of quality for a particular sampling plan.

6.2.3 AVERAGE OUTGOING QUALITY LIMIT

The relationship between the quality of the batch presented and the probably of acceptance based upon the quality of the sample leads to the creation of the **average outgoing quality** (AOQ). This average quality of the outgoing product or service is a result of the operation of sampling as shown in Figure 6.5.

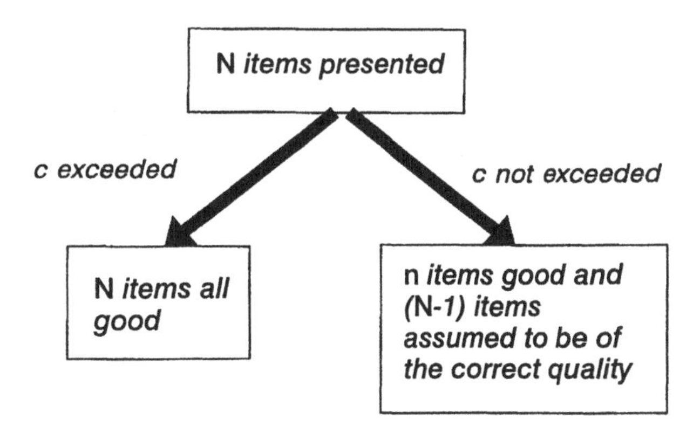

Figure 6.5 Operation of acceptance sampling.

The average outgoing quality is therefore given by:

$$\text{AOQ} = \text{Pa PD} \frac{(\text{N} - n)}{\text{N}}$$

where:

- Pa = probability of accepting batch with percentage defect PD;
- PD = defect percentage of the batch;
- N = batch size;
- n = sample size.

The relationship between the probability of acceptance and the batch quality (percentage defect level) as described by the typical OC curve indicates that the probability reduces as the defect level increases in a non-linear way. The result of this characteristic is that as the defect level in the batches increases, the AOQ experiences a maxima as shown in Figure 6.6.

The significance of this maxima, which is termed the average outgoing quality limit (AOQL), is:

- it represents the 'worst' case in terms of the average quality over a number of batches delivered to the customer and indicates the actual defect level in the batch that generates this worst case;
- it can be used as a 'guarantee' to the customer in that, irrespective of the actual production quality, the product or service quality will never be worse than the AOQL.

In practice the relationship shown in Figure 6.6 merely illustrates that as the quality of the product or service deteriorates then the probability that the batch will be accepted on the basis of the sampling plan also reduces. The AOQL is therefore a commonly used measure of the performance of a particular sampling plan.

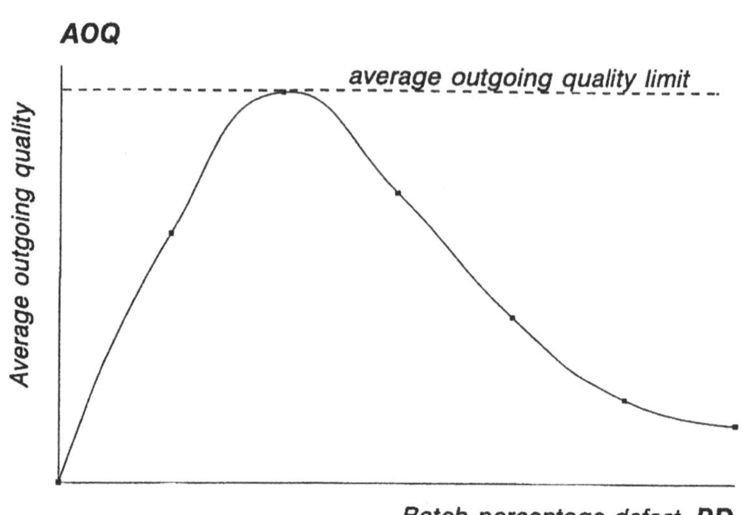

Figure 6.6 Average outgoing quality limit.

6.2.4 SUMMARY

- The operation of acceptance sampling is described using the operating characteristic curve which illustrates the relationship between the probability of accepting a batch and the batch quality level for a particular sampling plan.
- The average total number of items in a batch to be inspected is the sum of the sample size plus the extra inspection due to the producer's risk.
- The average outgoing quality reaches a limit which represents the maximum defect level which could be received by the customer.

6.3 Design of sampling plans

6.3.1 SAMPLE SIZE AND ACCEPTANCE NUMBER

In terms of designing a particular sampling plan the key questions are:

- how many should be sampled (n);
- how many defects are permissable in the sample (c).

For example, a service company attempting to evaluate customer satisfaction must decide how many customers to survey and, given a certain required quality level (say, fewer than 3% dissatisfied customers), how many of the customers in the survey are allowed to be dissatisfied.

The operating characteristic curve described above in section 6.2.1 is the relationship between the probability of acceptance and the defect level for a given sampling plan, that is a specific value of both n and c.

As the value of either n or c is changed then the OC curve changes and a family of curves are produced as shown in Figure 6.7.

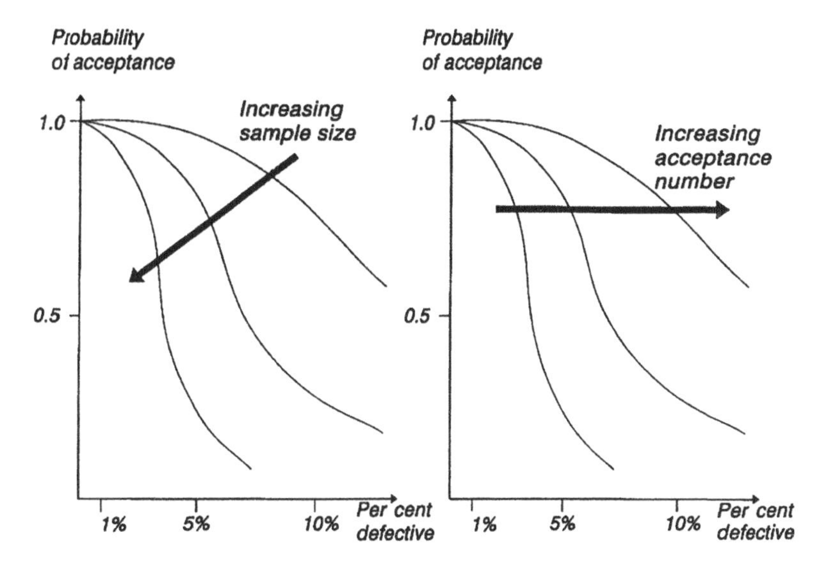

Figure 6.7 Variation of OC curves with values of sample size, n, and acceptance number, c.

There are two basic approaches to the design of sampling plans:

- using distribution tables such as the modified Thorndike chart;
- using national and international standards tables.

The first of theses techniques enables the sampling plan designer to work from first principles and so to design plans in line with the accepted levels of both producer's and consumer's risk. Where these risks cannot be established (for example, where a customer is unwilling to specify an agreed likelihood of receiving defective batches) then the application of published standards offers a reasonable business compromise. In other situations specific contractual requirements may invoke sampling plans to established standards and very often the requirements of ISO 9000 (Chapter 2) are met using plans derived from known standards.

6.3.2 THE DESIGN OF SAMPLING PLANS USING THE POISSON DISTRIBUTION

This approach to the design of sampling plans utilizes the unique relationship on the OC curve of the following two points:

- α and AQL;
- β and LQ.

The approach is therefore to 'fit' a curve to both of these points. This is done by relating the data on the modified Thorndike chart shown in Figure 6.3 to the points α, AQL and β, LQ. For the following batch of size N = 500, the agreed values of producers' and consumers' risks, for example, are:

α = 5% AQL = 1% defective

This is the probability of rejecting an acceptable batch, hence the probability of acceptance = $1 - \alpha = 95\%$.

β = 10% LQ = 5% defective

These four values uniquely define two points on the OC curve and therefore define a single curve and a unique sampling plan. The modified Thorndike chart is a plot of the relationship between the probability of occurrence and the average number of defects in the batch:

$$\text{where: Average number of defects} = \frac{\text{PD} \times n}{100}$$

To design the sampling plan using this approach the procedure is therefore:

1. Take one of the two known points (say, $1 - \alpha$, AQL) where the specific relationship between probability and defect level is known.
2. Read across the modified Thorndike chart at the known value of probability ($1 - \alpha$) for each of the acceptance values $c = 0$, $c = 1$, $c = 2$, $c = 3$, ..., and read off the corresponding values of PD $\times n_0/100$, PD $\times n_1/100$, PD $\times n_2/100$, PD $\times n_3/100$, ...
3. Repeat the procedure using the other known point (β, LQ) and for each of the values $c = 0$, $c = 1$, $c = 2$, etc., again determine the corresponding values of n_0, n_1, n_2, etc.
4. Compare the two sets of results and select the plan where the value of c and n are the same for both points.

This approach is illustrated for the example given in Table 6.1. Hence the plan that fits both points ($1 - \alpha$), AQL, and β, LQ, is one in which the sample size, n is 135 and the acceptance number, c is 3.

6.3.3 STANDARDS FOR THE DESIGN OF SAMPLING PLANS

The development of a set of standard sampling plans began during the Second World War with the ABC agreement between the American, British and Canadian governments.

In the UK this approach to the design of sampling plans is now embodied in the following published standard and its associated guide:

Table 6.1 Design of sampling plan using the modified Thorndike chart

Value of c	Average number of defects (PD × n/100)	Value of n
For α = 5% and AQL = 1%		
1	0.34	34
2	0.83	83
3	1.35	135
4	1.95	195
5	2.60	260
For β = 10% and LQ = 5%		
1	3.90	78
2	5.35	107
3	6.75	135
4	8.00	160
5	9.20	184

- BS 6001 *Specification for Sampling Procedures and Tables for Inspection by Attributes.*
- BS 6000 *Guide to the Use of BS 6001, Sampling Procedures and Tables for Inspection by Attributes.*

The international equivalents of these British Standards are ISO 2859–1, *Standard Procedures and Tables for Inspection by Attributes* and the American Military Standard MIL-STD 105E.

The approach used by these standards is to publish tables which list values of n and c for various types of plan, batch sizes, acceptable quality levels (AQL) and operational performance.

The basic procedure requires the following to be known:

- the batch or lot size, N;
- the acceptance quality level, AQL;
- the initial inspection level, normally level II.

From the batch size and the initial inspection level, Table I of the standard is used to determine the appropriate code letter. For example, consider a batch size of 500 with an AQL = 1% and an initial inspection level of II, Table I (shown in Figure 6.8) gives a code letter = H.

The simplest form of sampling involves taking a **single** sample from the batch and deciding whether to accept the batch on the basis of whether the number of defectives exceeds the value of c. Multiple sampling is also tabulated and is described below in section 6.4.

TABLE I—*Sample size code letters*

Lot or batch size			Special inspection levels				General inspection levels		
			S-1	S-2	S-3	S-4	I	II	III
2	to	8	A	A	A	A	A	A	B
9	to	15	A	A	A	A	A	B	C
16	to	25	A	A	B	B	B	C	D
26	to	50	A	B	B	C	C	D	E
51	to	90	B	B	C	C	C	E	F
91	to	150	B	B	C	D	D	F	G
151	to	280	B	C	D	E	E	G	H
281	to	500	B	C	D	E	F	H	J
501	to	1200	C	C	E	F	G	J	K
1201	to	3200	C	D	E	G	H	K	L
3201	to	10000	C	D	F	G	J	L	M
10001	to	35000	C	D	F	H	K	M	N
35001	to	150000	D	E	G	J	L	N	P
150001	to	500000	D	E	G	J	M	P	Q
500001	and	over	D	E	H	K	N	Q	R

CODE LETTERS

BS 6001 : 1972

Figure 6.8 Table I from BS 6001.*

TABLE II-A—Single sampling plans for normal inspection (Master table)

Sample size code letter	Sample size	Acceptable Quality Levels (normal inspection)																									
		0.010	0.015	0.025	0.040	0.065	0.10	0.15	0.25	0.40	0.65	1.0	1.5	2.5	4.0	6.5	10	15	25	40	65	100	150	250	400	650	1000
		Ac Re	Ac Re	Ac Re	Ac Re	Ac Re	Ac Re	Ac Re	Ac Re	Ac Re	Ac Re	Ac Re	Ac Re	Ac Re	Ac Re	Ac Re	Ac Re	Ac Re	Ac Re	Ac Re	Ac Re	Ac Re	Ac Re	Ac Re	Ac Re	Ac Re	Ac Re
A	2	↓	↓	↓	↓	↓	↓	↓	↓	↓	↓	↓	↓	↓	↓	0 1	↓	↓	1 2	2 3	3 4	5 6	7 8	10 11	14 15	21 22	30 31
B	3	↓	↓	↓	↓	↓	↓	↓	↓	↓	↓	↓	↓	↓	0 1	↑	↓	1 2	2 3	3 4	5 6	7 8	10 11	14 15	21 22	30 31	44 45
C	5	↓	↓	↓	↓	↓	↓	↓	↓	↓	↓	↓	↓	0 1	↑	↓	1 2	2 3	3 4	5 6	7 8	10 11	14 15	21 22	30 31	44 45	↑
D	8	↓	↓	↓	↓	↓	↓	↓	↓	↓	↓	↓	0 1	↑	↓	1 2	2 3	3 4	5 6	7 8	10 11	14 15	21 22	30 31	44 45	↑	↑
E	13	↓	↓	↓	↓	↓	↓	↓	↓	↓	↓	0 1	↑	↓	1 2	2 3	3 4	5 6	7 8	10 11	14 15	21 22	30 31	44 45	↑	↑	↑
F	20	↓	↓	↓	↓	↓	↓	↓	↓	↓	0 1	↑	↓	1 2	2 3	3 4	5 6	7 8	10 11	14 15	21 22	↑	↑	↑	↑	↑	↑
G	32	↓	↓	↓	↓	↓	↓	↓	↓	0 1	↑	↓	1 2	2 3	3 4	5 6	7 8	10 11	14 15	21 22	↑	↑	↑	↑	↑	↑	↑
H	50	↓	↓	↓	↓	↓	↓	↓	0 1	↑	↓	1 2	2 3	3 4	5 6	7 8	10 11	14 15	21 22	↑	↑	↑	↑	↑	↑	↑	↑
J	80	↓	↓	↓	↓	↓	↓	0 1	↑	↓	1 2	2 3	3 4	5 6	7 8	10 11	14 15	21 22	↑	↑	↑	↑	↑	↑	↑	↑	↑
K	125	↓	↓	↓	↓	↓	0 1	↑	↓	1 2	2 3	3 4	5 6	7 8	10 11	14 15	21 22	↑	↑	↑	↑	↑	↑	↑	↑	↑	↑
L	200	↓	↓	↓	↓	0 1	↑	↓	1 2	2 3	3 4	5 6	7 8	10 11	14 15	21 22	↑	↑	↑	↑	↑	↑	↑	↑	↑	↑	↑
M	315	↓	↓	↓	0 1	↑	↓	1 2	2 3	3 4	5 6	7 8	10 11	14 15	21 22	↑	↑	↑	↑	↑	↑	↑	↑	↑	↑	↑	↑
N	500	↓	↓	0 1	↑	↓	1 2	2 3	3 4	5 6	7 8	10 11	14 15	21 22	↑	↑	↑	↑	↑	↑	↑	↑	↑	↑	↑	↑	↑
P	800	↓	0 1	↑	↓	1 2	2 3	3 4	5 6	7 8	10 11	14 15	21 22	↑	↑	↑	↑	↑	↑	↑	↑	↑	↑	↑	↑	↑	↑
Q	1250	0 1	↑	↓	1 2	2 3	3 4	5 6	7 8	10 11	14 15	21 22	↑	↑	↑	↑	↑	↑	↑	↑	↑	↑	↑	↑	↑	↑	↑
R	2000	↑	↑	1 2	2 3	3 4	5 6	7 8	10 11	14 15	21 22	↑	↑	↑	↑	↑	↑	↑	↑	↑	↑	↑	↑	↑	↑	↑	↑

↓ = Use first sampling plan below arrow. If sample size equals, or exceeds, lot or batch size, do 100 percent inspection.
↑ = Use first sampling plan above arrow.
Ac = Acceptance number.
Re = Rejection number.

Figure 6.9 Single sampling for normal inspection from BS 6001.*

To determine the appropriate single sampling plan, the corresponding table should be referred to. For the code letter H from the example above the **normal** inspection table for a single sampling plan shown in Figure 6.9 reveals the correct plan to be a sample size of 50 and an acceptance number of 1 (reject number of 2).

If the table does not indicate a value of c directly, then the arrows indicate the direction to follow to determine the value of c to be used. It should also be noticed that the values of AQL increase to a value of 1000 which clearly does not correspond to a percentage defective. The correct interpretation of these levels is that from 0.1% to 10% corresponds to AQLs expressed as a percentage of defectives in the complete batch or lot, whereas values over 10% are used where the number of defectives are expressed per 100 units (one unit potentially having more than one defective, for example the number of scratches in paintwork). Finally, if two out of five batches or lots are rejected (the number of defects found in the sample exceeding the acceptance number, c) then **tightened** inspection is adopted which is displayed as a different table with lower values of c and larger values of n than the corresponding **normal** plan. If ten consecutive batches are accepted the normal sampling is resumed. If, however, under normal sampling five consecutive batches are approved then the sampling can revert to **reduced** sampling (displayed as alternative tables having smaller samples and greater acceptance numbers) until a batch is rejected, at which point normal sampling is resumed. This process of adapting the sampling plan is shown in Figure 6.10. Such tables are easily used and eliminate the need for curve fitting as required by the Poisson techniques.

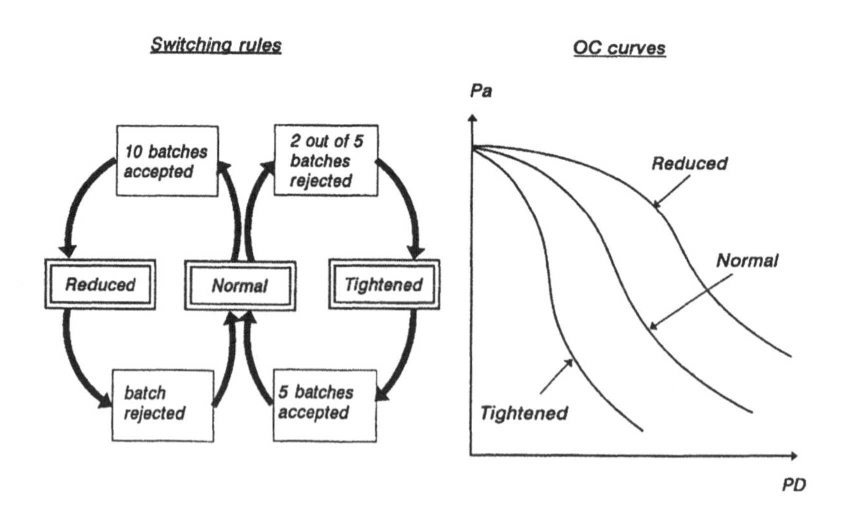

Figure 6.10 The application of normal, tightened and reduced inspection.

BS 6001 also contains full operating characteristic curves for the sampling plans contained within.

6.3.4 SUMMARY

- The design of a sampling plan involves the selection of the sample size and the acceptable number of defects permitted in the sample.
- Sampling plans can be designed using the Poisson distribution and determining common values of sample size and acceptance number from the modified Thorndike chart.
- Sampling plans can also be readily determined from standards tables such as those found in BS 6001 or ISO 2859–1.

6.4 Multiple and sequential sampling

6.4.1 THE ROLE OF MULTIPLE AND SEQUENTIAL SAMPLING PLANS

In many practical situations, the taking of a single sample is not an appropriate approach to acceptance sampling.

Where the sampling may be disruptive, destructive or expensive then it may be more appropriate to take a number of smaller samples than the single sample size, n, if a decision could be made on the basis of the reduced cumulative samples. This taking of two or more smaller samples rather than a single larger sample is termed **multiple sampling**. This approach to sampling is useful where the process under inspection operates either well within the AQL or well outside the LTPD – in other words the process is either very good or very bad and rarely indifferent. In this situation a multiple sampling regime offers the earlier opportunity to make a decision to accept or reject the batch on, say, the first or second (smaller) samples. The taking of more than one sample, however, may be in itself inconvenient or disruptive and therefore multiple sampling is generally only used where the variation in quality level in the batch offers a good opportunity to make a decision on a smaller total sample size. The operation of multiple acceptance sampling is described below in section 6.4.2.

Sequential sampling involves the taking of individual samples throughout the production of the batch rather than waiting until the batch is completed and then taking a single (or multiple sample). This approach has obvious advantages in situations where:

- it is important to gain knowledge of the quality of the batch at an early stage in the production due to, say, expensive or time-consuming corrective actions;
- the batch is produced over a long period of time which may make it very inconvenient to wait until the end of the production run before finding the process is defective and unacceptable.

The design and operation of sequential sampling plans is described below in section 6.4.3.

In both manufacturing and service sector applications of acceptance sampling it is important to select the most appropriate approach to sampling to ensure that the overall cost of assessment is minimized and the benefits maximized.

6.4.2 THE DESIGN OF MULTIPLE SAMPLING PLANS

There are two forms of sampling plans in which more than a single sample is taken:

- **double sampling plans** in which two samples are potentially taken;
- **multiple sampling plans** where up to seven samples may eventually be taken.

The most common approach to the design of multiple sampling plans is to use standard tables such as those given in BS 6001/ISO 2859–1 and MIL-STD 105E. The main difference between the design of a multiple sampling plan compared to a single plan is that in addition to having a series of sample sizes, the plans quote both an accept number and a reject number which only give a definite outcome at the final sample. The basic procedure for the operation of multiple sampling is as follows:

1. Take the initial sample and determine the number of defectives present.
2. If the number of defectives is less than the accept number (Ac) then the whole batch should be accepted, or if the number of defectives is greater than the reject number (Re) then the whole batch is rejected and subjected to further 100% inspection.
3. If the number of defects in the initial sample falls between Ac and Re then a second sample must be taken and the number of defects found in the second sample added to the number found in the first.
4. The total number of defects should then be compared with the accept number and reject number for the second sample and again a decision is made or a further sample taken.
5. At the final sample the difference between the Ac and Re is equal to one and therefore a decision to accept or reject is forthcoming on the basis of the cumulative number of defects found in all the samples.

The tables for the selection of double and multiple sampling plans are shown in Figure 6.11.

For the example used in the design of a single sampling plan in section 6.3.3 where N = 500, AQL = 1% and code letter = H, from the data shown in Figure 6.11 the appropriate double sampling plan would be:

TABLE III-A—Double sampling plans for normal inspection (Master table)

(See 9.4 and 9.5)

Sample size code letter	Sample	Sample size	Cumulative sample size	Acceptable Quality Levels (normal inspection)																									
				0.010	0.015	0.025	0.040	0.065	0.10	0.15	0.25	0.40	0.65	1.0	1.5	2.5	4.0	6.5	10	15	25	40	65	100	150	250	400	650	1000
				Ac Re	Ac Re	Ac Re	Ac Re	Ac Re	Ac Re	Ac Re	Ac Re	Ac Re	Ac Re	Ac Re	Ac Re	Ac Re	Ac Re	Ac Re	Ac Re	Ac Re	Ac Re	Ac Re	Ac Re	Ac Re	Ac Re	Ac Re	Ac Re	Ac Re	Ac Re
A				↓	↓	↓	↓	↓	↓	↓	↓	↓	↓	↓	↓	↓	↓	•	↓	↓	•	•	•	•	•	•	•	•	•
B	First	2	2	↓	↓	↓	↓	↓	↓	↓	↓	↓	↓	↓	↓	↓	•	↑	↓	0 2	0 3	1 4	2 5	3 7	5 9	7 11	11 16	17 22	25 31
	Second	2	4																	1 2	3 4	4 5	6 7	8 9	12 13	18 19	26 27	37 38	56 57
C	First	3	3	↓	↓	↓	↓	↓	↓	↓	↓	↓	↓	↓	↓	•	↑	↓	0 2	0 3	1 4	2 5	3 7	5 9	7 11	11 16	17 22	25 31	↑
	Second	3	6																1 2	3 4	4 5	6 7	8 9	12 13	18 19	26 27	37 38	56 57	
D	First	5	5	↓	↓	↓	↓	↓	↓	↓	↓	↓	↓	↓	•	↑	↓	0 2	0 3	1 4	2 5	3 7	5 9	7 11	11 16	17 22	25 31	↑	↑
	Second	5	10															1 2	3 4	4 5	6 7	8 9	12 13	18 19	26 27	37 38	56 57		
E	First	8	8	↓	↓	↓	↓	↓	↓	↓	↓	↓	↓	•	↑	↓	0 2	0 3	1 4	2 5	3 7	5 9	7 11	11 16	17 22	25 31	↑	↑	↑
	Second	8	16														1 2	3 4	4 5	6 7	8 9	12 13	18 19	26 27	37 38	56 57			
F	First	13	13	↓	↓	↓	↓	↓	↓	↓	↓	↓	•	↑	↓	0 2	0 3	1 4	2 5	3 7	5 9	7 11	11 16	↑	↑	↑	↑	↑	↑
	Second	13	26													1 2	3 4	4 5	6 7	8 9	12 13	18 19	26 27						
G	First	20	20	↓	↓	↓	↓	↓	↓	↓	↓	•	↑	↓	0 2	0 3	1 4	2 5	3 7	5 9	7 11	11 16	↑	↑	↑	↑	↑	↑	↑
	Second	20	40											1 2	3 4	4 5	6 7	8 9	12 13	18 19	26 27								
H	First	32	32	↓	↓	↓	↓	↓	↓	↓	•	↑	↓	0 2	0 3	1 4	2 5	3 7	5 9	7 11	11 16	↑	↑	↑	↑	↑	↑	↑	↑
	Second	32	64											1 2	3 4	4 5	6 7	8 9	12 13	18 19	26 27								
J	First	50	50	↓	↓	↓	↓	↓	↓	•	↑	↓	0 2	0 3	1 4	2 5	3 7	5 9	7 11	11 16	↑	↑	↑	↑	↑	↑	↑	↑	↑
	Second	50	100										1 2	3 4	4 5	6 7	8 9	12 13	18 19	26 27									
K	First	80	80	↓	↓	↓	↓	↓	•	↑	↓	0 2	0 3	1 4	2 5	3 7	5 9	7 11	11 16	↑	↑	↑	↑	↑	↑	↑	↑	↑	↑
	Second	80	160									1 2	3 4	4 5	6 7	8 9	12 13	18 19	26 27										
L	First	125	125	↓	↓	↓	↓	•	↑	↓	0 2	0 3	1 4	2 5	3 7	5 9	7 11	11 16	↑	↑	↑	↑	↑	↑	↑	↑	↑	↑	↑
	Second	125	250								1 2	3 4	4 5	6 7	8 9	12 13	18 19	26 27											
M	First	200	200	↓	↓	↓	•	↑	↓	0 2	0 3	1 4	2 5	3 7	5 9	7 11	11 16	↑	↑	↑	↑	↑	↑	↑	↑	↑	↑	↑	↑
	Second	200	400							1 2	3 4	4 5	6 7	8 9	12 13	18 19	26 27												
N	First	315	315	↓	↓	•	↑	↓	0 2	0 3	1 4	2 5	3 7	5 9	7 11	11 16	↑	↑	↑	↑	↑	↑	↑	↑	↑	↑	↑	↑	↑
	Second	315	630						1 2	3 4	4 5	6 7	8 9	12 13	18 19	26 27													
P	First	500	500	↓	•	↑	↓	0 2	0 3	1 4	2 5	3 7	5 9	7 11	11 16	↑	↑	↑	↑	↑	↑	↑	↑	↑	↑	↑	↑	↑	↑
	Second	500	1000					1 2	3 4	4 5	6 7	8 9	12 13	18 19	26 27														
Q	First	800	800	•	↑	↓	0 2	0 3	1 4	2 5	3 7	5 9	7 11	11 16	↑	↑	↑	↑	↑	↑	↑	↑	↑	↑	↑	↑	↑	↑	↑
	Second	800	1600				1 2	3 4	4 5	6 7	8 9	12 13	18 19	26 27															
R	First	1250	1250	↑	↑	0 2	0 3	1 4	2 5	3 7	5 9	7 11	11 16	↑	↑	↑	↑	↑	↑	↑	↑	↑	↑	↑	↑	↑	↑	↑	↑
	Second	1250	2500			1 2	3 4	4 5	6 7	8 9	12 13	18 19	26 27																

↓ = Use first sampling plan below arrow. If sample size equals or exceeds lot or batch size, do 100 percent inspection.
↑ = Use first sampling plan above arrow.
Ac = Acceptance number
Re = Rejection number
• = Use corresponding single sampling plan (or alternatively, use double sampling plan below, where available).

DOUBLE NORMAL

Figure 6.11 Double and multiple sampling plans from BS 6001.*

MULTIPLE
NORMAL

TABLE IV-A—Multiple sampling plans for normal inspection (Master table)

(See 9.4 and 9.5)

Sample size code letter	Sample	Sample size	Cumulative sample size	Acceptable Quality Levels (normal inspection)																																																			
				0.010		0.015		0.025		0.040		0.065		0.10		0.15		0.25		0.40		0.65		1.0		1.5		2.5		4.0		6.5		10		15		25		40		65		100		150		250		400		650		1000	
				Ac	Re	Ac	Re	Ac	Re	Ac	Re	Ac	Re	Ac	Re	Ac	Re	Ac	Re	Ac	Re	Ac	Re	Ac	Re	Ac	Re	Ac	Re	Ac	Re	Ac	Re	Ac	Re	Ac	Re	Ac	Re	Ac	Re	Ac	Re	Ac	Re	Ac	Re	Ac	Re	Ac	Re	Ac	Re	Ac	Re
A				↓		↓		↓		↓		↓		↓		↓		↓		↓		↓		↓		↓		↓		↓		•		↓		↓		•		•		•		•		•		•		•		•		•	
B																												•		•		↑				++		++		++		++		++		++		++		++		++		++	
C																														↑		↓		++		++		++		++		++		++		++		++		++		++		↑	
D	First	2	2																							•		↑		↓		*	2	*	2	*	3	*	4	0	4	0	5	1	7	2	9	4	12	6	16	↑			
	Second	2	4																													*	2	0	3	0	3	1	5	1	6	3	8	4	10	7	14	11	19	17	27				
	Third	2	6																													0	2	0	3	1	4	2	6	3	8	6	10	8	13	13	19	19	27	29	39				
	Fourth	2	8																													0	3	1	4	2	5	3	7	5	10	8	13	12	17	19	25	27	34	40	49				
	Fifth	2	10																													1	3	2	4	3	6	5	8	7	11	11	15	17	20	25	29	36	40	53	58				
	Sixth	2	12																													1	3	3	5	4	6	7	9	10	12	14	17	21	23	31	33	45	47	65	68				
	Seventh	2	14																													2	3	4	5	6	7	9	10	13	14	18	19	25	26	37	38	53	54	77	78				
E	First	3	3																					•		↑		↓		*	2	*	2	*	3	*	4	0	4	0	5	1	7	2	9	4	12	6	16	↑					
	Second	3	6																											*	2	0	3	0	3	1	5	1	6	3	8	4	10	7	14	11	19	17	27						
	Third	3	9																											0	2	0	3	1	4	2	6	3	8	6	10	8	13	13	19	19	27	29	39						
	Fourth	3	12																											0	3	1	4	2	5	3	7	5	10	8	13	12	17	19	25	27	34	40	49						
	Fifth	3	15																											1	3	2	4	3	6	5	8	7	11	11	15	17	20	25	29	36	40	53	58						
	Sixth	3	18																											1	3	3	5	4	6	7	9	10	12	14	17	21	23	31	33	45	47	65	68						
	Seventh	3	21																											2	3	4	5	6	7	9	10	13	14	18	19	25	26	37	38	53	54	77	78						
F	First	5	5																			•		↑		↓		*	2	*	2	*	3	*	4	0	4	0	5	1	7	2	9	↑		↑		↑							
	Second	5	10																									*	2	0	3	0	3	1	5	1	6	3	8	4	10	7	14												
	Third	5	15																									0	2	0	3	1	4	2	6	3	8	6	10	8	13	13	19												
	Fourth	5	20																									0	3	1	4	2	5	3	7	5	10	8	13	12	17	19	25												
	Fifth	5	25																									1	3	2	4	3	6	5	8	7	11	11	15	17	20	25	29												
	Sixth	5	30																									1	3	3	5	4	6	7	9	10	12	14	17	21	23	31	33												
	Seventh	5	35																									2	3	4	5	6	7	9	10	13	14	18	19	25	26	37	38												
G	First	8	8																	•		↑		↓		*	2	*	2	*	3	*	4	0	4	0	5	1	7	2	9	↑													
	Second	8	16																							*	2	0	3	0	3	1	5	1	6	3	8	4	10	7	14														
	Third	8	24																							0	2	0	3	1	4	2	6	3	8	6	10	8	13	13	19														
	Fourth	8	32																							0	3	1	4	2	5	3	7	5	10	8	13	12	17	19	25														
	Fifth	8	40																							1	3	2	4	3	6	5	8	7	11	11	15	17	20	25	29														
	Sixth	8	48																							1	3	3	5	4	6	7	9	10	12	14	17	21	23	31	33														
	Seventh	8	56																							2	3	4	5	6	7	9	10	13	14	18	19	25	26	37	38														
H	First	13	13															•		↑		↓		*	2	*	2	*	3	*	4	0	4	0	5	1	7	2	9	↑															
	Second	13	26																					*	2	0	3	0	3	1	5	1	6	3	8	4	10	7	14																
	Third	13	39																					0	2	0	3	1	4	2	6	3	8	6	10	8	13	13	19																
	Fourth	13	52																					0	3	1	4	2	5	3	7	5	10	8	13	12	17	19	25																
	Fifth	13	65																					1	3	2	4	3	6	5	8	7	11	11	15	17	20	25	29																
	Sixth	13	78																					1	3	3	5	4	6	7	9	10	12	14	17	21	23	31	33																
	Seventh	13	91																					2	3	4	5	6	7	9	10	13	14	18	19	25	26	37	38																
J	First	20	20													•		↑		↓		*	2	*	2	*	3	*	4	0	4	0	5	1	7	2	9	↑																	
	Second	20	40																			*	2	0	3	0	3	1	5	1	6	3	8	4	10	7	14																		
	Third	20	60																			0	2	0	3	1	4	2	6	3	8	6	10	8	13	13	19																		
	Fourth	20	80																			0	3	1	4	2	5	3	7	5	10	8	13	12	17	19	25																		
	Fifth	20	100																			1	3	2	4	3	6	5	8	7	11	11	15	17	20	25	29																		
	Sixth	20	120																			1	3	3	5	4	6	7	9	10	12	14	17	21	23	31	33																		
	Seventh	20	140																			2	3	4	5	6	7	9	10	13	14	18	19	25	26	37	38																		

↓ = Use first sampling plan below arrow (refer to continuation of table on following page, when necessary). If sample size equals or exceeds lot or batch size, do 100 percent inspection.
↑ = Use first sampling plan above arrow.
Ac = Acceptance number.
Re = Rejection number.
• = Use corresponding single sampling plan (or alternatively, use multiple sampling plan below, where available).
++ = Use corresponding double sampling plan (or alternatively, use multiple sampling plan below, where available).
* = Acceptance not permitted at this sample size.

Figure 6.11 *Contd*

First sample = 32
Accept number = 0 Reject number 2

If the number of defects in the first sample is found to be 1 then a second sample is taken:

Second sample = 32
Accept number = 1 Reject number 2

If a further defect is found in the second sample then the total number of defects found in both samples would be 2 and therefore the batch would be rejected.

It should be noted that in comparison with the corresponding single sampling plan quoted in section 6.4.2 where the single sample size was 50, the total double sample size is 64. Double sampling does, however, offer the opportunity of making a decision after the first sample of only 32 items.

Figure 6.11 also shows that the corresponding multiple sample plan has a minimum (first) sample size of 13 and a maximum sample of 91 which again has to be compared with the single sample size of 50.

These are the factors to be considered when selecting double and multiple plans over single sampling.

6.4.3 THE DESIGN OF SEQUENTIAL SAMPLING PLANS

Sequential sampling is an important technique in situations where it is beneficial to monitor the ongoing quality of production. This approach is appropriate in manufacturing situations where the batch is produced over a long period of time or in service situations, for example where customers are surveyed sequentially to monitor the development of a market preference.

As with single sampling plan design there are two approaches which can be adopted:

- sequential sampling plan design using the values of ($1 - \alpha$, AQL) and (β, LQ);
- sequential sampling plan design using published standards.

Both of these are graphical techniques in which the sequence in which the individual samples pass or fail is critical in terms of whether the batch or lot is accepted or rejected.

The first of these techniques offers the advantage of the Poisson distribution approach to single sampling plan design in that the plan can be developed from first principles providing the relationships pertaining to both producer's and consumer's risks are known. The procedure is to construct a graph of the relationship between the cumulative number of defectives found and sequential sample number. Onto this graph are drawn two lines which represent the fail region and the pass region, as shown in Figure 6.12.

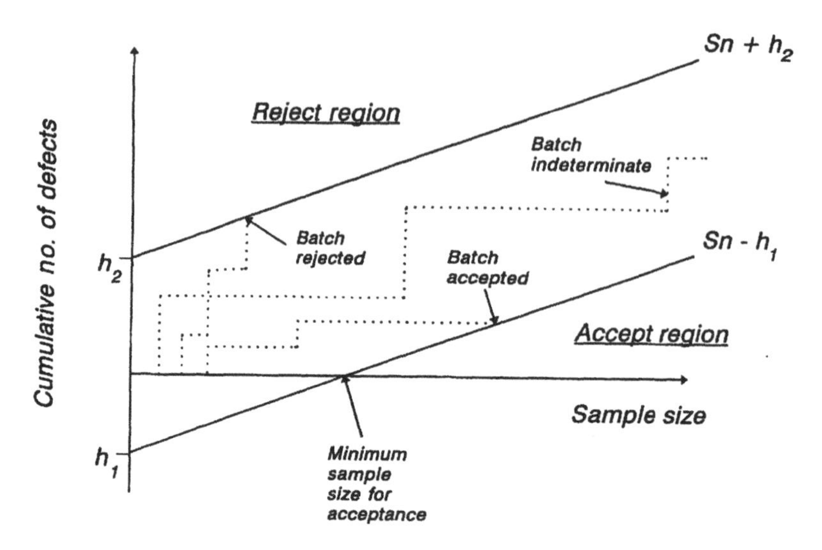

Figure 6.12 Sequential sampling using producer's and consumer's risks.

These lines are given by:

$$Sn + h_1 \qquad \text{(Fail region)}$$

and:

$$Sn - h_2 \qquad \text{(Pass region)}$$

where:

$$h_1 = \frac{\log_{10}(\frac{1-\alpha}{\beta})}{\log_{10}[\frac{p_2}{p_1}(\frac{1-p_1}{1-p_2})]}, \quad h_2 = \frac{\log_{10}(\frac{1-\beta}{\alpha})}{\log_{10}[\frac{p_2}{p_1}(\frac{1-p_1}{1-p_2})]}$$

also

$$s = \frac{\log_{10}(\frac{1-p_1}{1-p_2})}{\log_{10}[\frac{p_2}{p_1}(\frac{1-p_1}{1-p_2})]}$$

where: S = slope of the line; n = the sequential sample number;
h_1 = fail intercept; p_1 = AQL;
h_2 = pass intercept; p_2 = LQ.

The three examples shown in Figure 6.12 illustrate each of the possible outcomes of the sequential sampling plan:

- a 'pass' batch in which the cumulative number of defects line passes into the passed region;
- a 'fail' batch in which the cumulative number of defects line passes into the fail region;

- an 'indeterminate' batch in which the cumulative number of defects line remains within the zone between the pass and fail regions. In this case a decision cannot be made to accept or reject the batch on the basis of sequential sampling and an equivalent multiple plan (section 6.4.2) would need to be selected.

With the sequential sampling arrangement shown in Figure 6.12 it can be seen that a 'minimum sample size' exists at the point at which the pass region crosses the y axis (zero cumulative defects). If there are no defects found in the sequentially inspected sample then this point represents the smallest sample size needed to accept a batch.

The alternative approach to the design of sequential sampling plans is to adopt a plan published in national or international standards such as BS 6001. The sequential sampling plans specified in the standards are normally defined in terms of:

- the handicap, H;
- the penalty, b.

The procedure is again to construct a chart with the fail region set at zero and the pass region set at a value of twice the handicap. The sampling commences with a 'score' set equal to the handicap and each time a sample is tested and passed then the score is incremented by 1. If however, a sample is failed then the score is reduced by the value of the penalty. The batch is therefore accepted if the value of the score reaches a value of 2H or rejected if the score falls to zero, as illustrated in Figure 6.13.

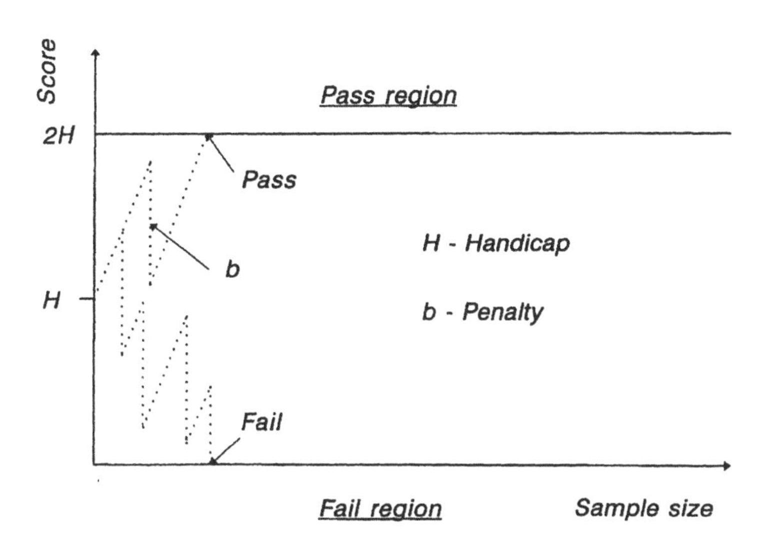

Figure 6.13 Sequential sampling using BS 6001.

Again, if the sequential sampling is indeterminate then the corresponding multiple sampling plan from BS 6001 is imposed.

6.4.4 SUMMARY

- Multiple and sequential sampling provide an alternative to single sampling plans and can offer the advantages of reduced inspection and an earlier decision on acceptance or rejection.
- Multiple sampling plans from BS 6001/ISO 2859–1 have accept and reject numbers associated with each of the samples defined, and if a decision cannot be made a further sample is taken and the number of defects accumulated until conclusion.
- Sequential sampling utilizes a graphical approach developed either from the producer's/consumer's risk or from BS 6001.

Acknowledgement

* Extracts from BS6001: Part 1: 1991 are reproduced with the permission of BSI. Complete copies can be obtained by post from BSI Customer Services, 389 Chiswick High Road, London, W4 4AL, UK; tel. 0181-996 7000.

7 Statistical process control

7.1 The control of process variability

7.1.1 PROCESS VARIABILITY AND QUALITY DEVELOPMENT

> The statistical control of their manufacturing processes has been the key to Japan's dominance of world markets.

This quotation represents a commonly held viewpoint during the 1970s and 1980s in Western countries who began to import manufactured goods from Japan which were seen to be better quality and to be more reliable. Certainly Japan adopted the application of statistical methods to the management of quality rather more vigorously and rather earlier than the West, but the really key issue was in making the connection between the quality of the product or service and the variability inherent in the process responsible. It is ironic that many of the statistical techniques were pioneered in the West, and in particular in the telecommunications industry in the USA, where statisticians such as Edwards Deming and Walter Shewhart began to promote the benefits of applying statistical controls to manufacturing processes.

The reasons that these techniques were more readily adopted in Japan rather than in Western manufacturing industries after the Second World War are varied but certainly the application of these techniques assisted quality development in Japan.

In terms of developing the way in which quality is managed within an organization **statistical process control (SPC)** offers two important insights:

- improving the control of variation improves the probability that the product or service will conform to specification;
- there is a distinction between the variation present in the process and the tolerances which are allowed and which are derived from the specification.

As companies mature, SPC is increasingly used as an improvement methodology to improve the performance of the product through the reduction in variability. As companies mature yet further and develop a prevention orientation, the inherent variability of the process is 'built in' to the design of the product or service and the tolerances or specifications are set accordingly.

The implementation of SPC can also have a considerable cultural impact upon an organization in terms of:

- **understanding**, as employees begin to fundamentally appreciate and measure the behaviour of the processes with which they work;
- **involvement**, as employees become an effective part of the controlling mechanism and not simply observers of troublesome variation;
- **improvement**, as troubleshooting and process diagnostics become more systematic, effective and hence rewarding.

If processes did not exhibit variation then quality problems would not arise. It is, however, due to the fact that processes do suffer variation that there is a need to understand and control that variation. What SPC attempts to do is to distinguish between 'normal' and 'abnormal' variation which is a critical distinction in many business processes. Understanding when to intervene and also when not to intervene is very much part of a maturing approach to the management of quality.

7.1.2 PROCESS CONTROLS AND PROCESS COSTS

The economic importance of the effective control of variation is highlighted when the cost model, which assumes that the cost of non-conformance commences at the tolerance limits, is abandoned in favour of the view that costs increase as the process moves away from the target value (as discussed in Chapter 3). The inherent cost of the variability of the process manifests itself particularly where components are combined together to form systems and the collective effects of poor reliability multiply. Indeed it was the pressure to produce complex manufactured systems such as telephone exchanges that prompted the original work on the application of SPC. Until the variability of component parts is reduced (and hence reliability increased) then the manufacture of complex systems which are reliable becomes an impossibility. The main costs associated with the implementation and operation of SPC are shown in Figure 7.1 together with the life-cycle costs curves exhibited by many manufactured products.

When the external cost benefits are considered then the case for implementing SPC externally (with suppliers) as well as internally is extremely powerful and has led many major international organizations (such as the Ford Motor Company) to make the demonstration of effective process control (in addition to product control) a prerequisite for suppliers. The added confidence which is derived from knowing that suppliers have not only monitored the final quality of the product or service but also the manufacturing process helps reduce both incoming inspection costs and subsequent problems with supplied materials or services.

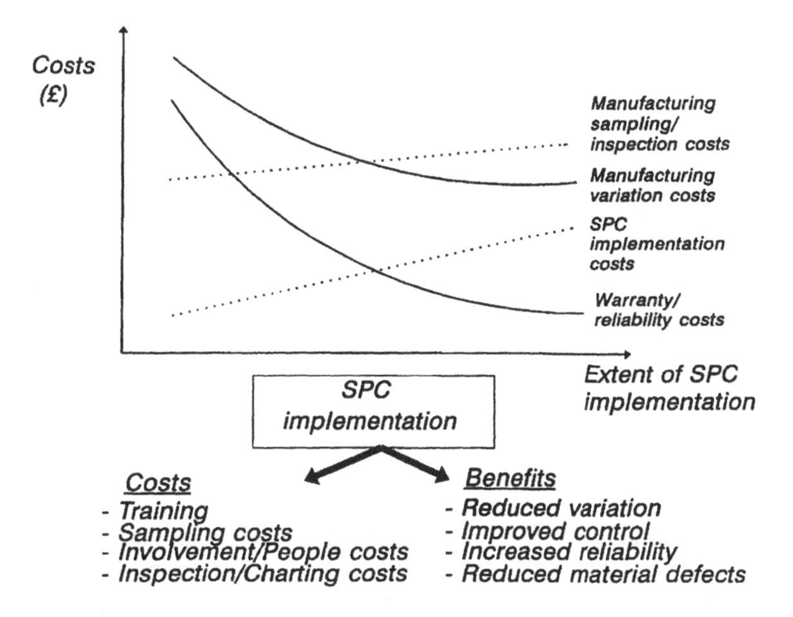

Figure 7.1 SPC implementation and lifecycle costs.

7.1.3 THE APPLICATION OF STATISTICAL PROCESS CONTROL

In many sectors of industry, the application of SPC has become a standard practice throughout the supply chain. Certainly in many of the mass production industries such as motor car or consumer electronics manufacture, SPC is applied extensively within the organizations and also encouraged within supplier companies.

The benefit of large (often multinational) organizations encouraging their suppliers to use SPC and to demonstrate the capability of their processes has been the promotion and widespread application of the statistical tools and techniques. The disadvantage of this customer imposed requirement has been the rather narrow focus upon the techniques, such as control charts, rather than upon the philosophy of understanding and improving (reducing) process variation.

The operation of process control charts as described below in section 7.5 is only one element of the application of SPC. Many of the benefits of using a systematic and statistical approach to analysing why a particular process varies are derived well before control charts are produced. Indeed very often improved process understanding leading to improved process capability can reduce or even eliminate the need for process control charts.

The application of SPC commences with developing an understanding of the process, identifying the main process parameters and defining the control mechanisms available. The technique which is widely used during this initial diagnostic phase is **failure mode effects and analysis (FMEA)**. This technique helps to identify which are the critical characteristics of the process, which require further, more detailed process analysis.

The second stage involves the collection and analysis of the process data to determine what is the **normal variation**. The process has to be bounded to include the inherent variability and to exclude the special causes of variation. This determining of 'what is normal?' requires a fundamental understanding of the process. The process data is also evaluated during this analysis phase to check whether the distribution is normal.

The next stage in the application of SPC involves the analysis of the critical characteristics to determine the relationship between the natural variation of the process and the allowable specification limits. This relationship is termed the **process capability** and is used to evaluate whether the process variability is acceptable.

Finally, having identified the critical characteristics of the process, having analysed the normal variation and having evaluated the capability of the process, then the ongoing control of the process can be monitored using control charts. The use of control charts therefore represents the final stage in the application of SPC, as illustrated in Figure 7.2.

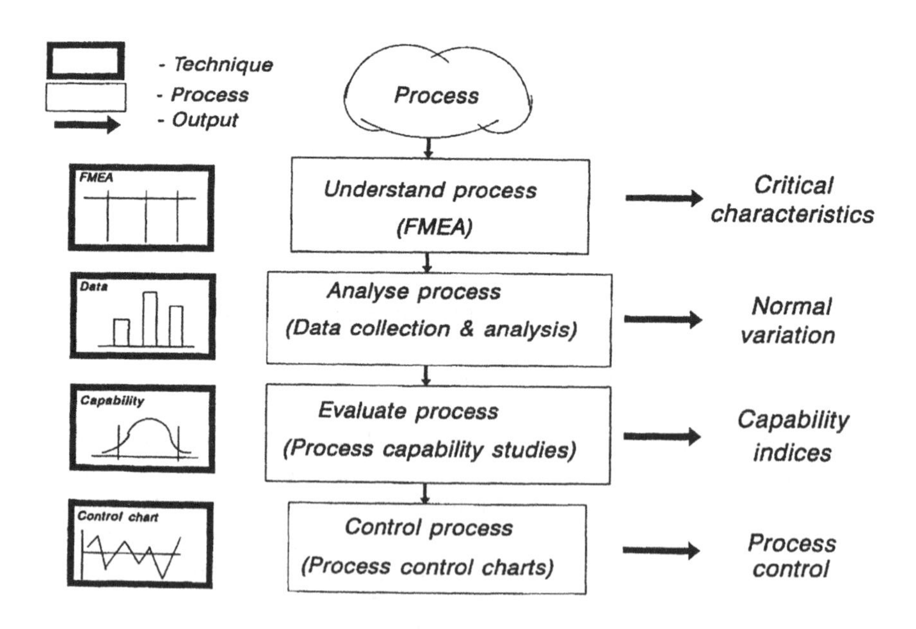

Figure 7.2 Stages in the application of statistical process control.

7.1.4 SUMMARY

- The control of process variation is an important element of quality development.
- The implementation of statistical process control can reduce the overall process cost.
- The stages in the implementation of statistical process control involve the identification of process failure modes, the analysis of the process data, the evaluation of the process capability and the charting of process performance.

7.2 Failure mode effects and analysis and critical characteristics

7.2.1 THE ROLE OF FMEA IN SPC

Having understood the importance of the relationship between quality and variability and hence the need for SPC, the question inevitably arises: 'What is it we should measure and control?' The basic guidelines for what constitutes a good measure are:

- measures should be readily obtained and relevant to the customer;
- measures should be quantitative rather than qualitative;
- measure should be timely and accurate.

The measures used to assess the quality of a product or service relate of course to the requirements of the specification and also in many instances develop over time within the customer–supplier relationship. When considering what to measure and control in an ongoing sense the variability of the parameter needs to be considered in addition to the importance of the parameter to the customer. When a new product or service is to be launched it is often extremely difficult to decide what to measure in order to control the ongoing conformance to specification. Very often aspects of the specification are subjective, take lengthy analysis or are difficult to obtain and therefore are not conveniently measurable.

In answer to this problem of having to arbitrarily decide which parameters should be monitored and controlled, the technique of failure mode effects and analysis (FMEA) was developed. There are two forms of FMEA.

- **design FMEA** which involves the analysis of the potential failures of a new product or service;
- **process FMEA** which involves failure analysis of the process or processes in the manufacture or provision of the product or service.

The role of the design FMEA is described in section 9.3. and is primarily used as a prevention orientated technique to improve the inherent reliability of a product or service.

The process FMEA is used primarily to identify areas of criticality to control and to emphasize the design of inherently more reliable processes. The preparation of a process FMEA should always be the starting point in the application of SPC. The FMEA assists in the understanding of the process and in the pinpointing of the critical areas to control.

7.2.2 PREPARING A FAILURE MODE AND EFFECTS ANALYSIS

The preparation of the FMEA should always be undertaken by a team of people drawn from all the relevant areas of the organization bringing the appropriate information and insights, as illustrated in Figure 7.3.

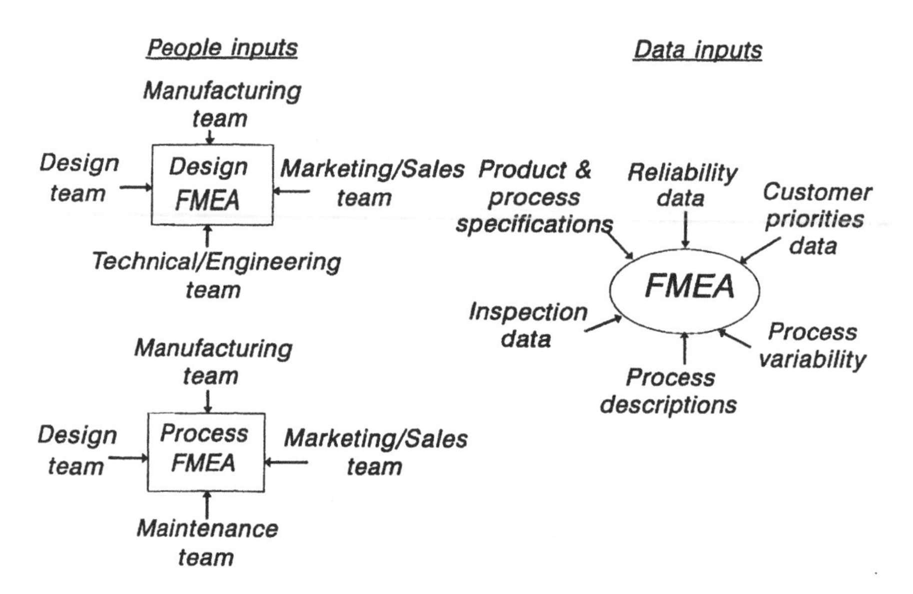

Figure 7.3 Typical inputs for the preparation of the FMEA.

The basic stages in the preparation of the FMEA are as follows.

- **Stage 1.** Identify the primary components or elements of the product or process and obtain agreement on the purpose.
- **Stage 2.** Identify the various modes of failure that the individual process components/elements may suffer (each component or element may have a number of failure modes).
- **Stage 3.** Identify the causes of failure for each of the failure modes identified in Stage 2.

- **Stage 4.** Identify the effects (both internally and to the customer) of each of the failure modes in terms of its impact upon the performance of the product or process.
- **Stage 5.** Identify the existing/proposed controls which are employed to monitor or prevent the mode of failure.
- **Stage 6.** Assess each of the failure modes in terms of the combined severity, occurrence and ability to detect the failure.
- **Stage 7.** From the relative priorities established in Stage 6, identify corrective actions to improve the most significant failure modes.
- **Stage 8.** Review the failure assessment identified in Stage 6 in the light of the corrective actions proposed in Stage 7.

The two key activities in the preparation of the FMEA are:

- the systematic identification of failure in terms of cause, effects and controls;
- the assessment of the combined severity, occurrence and detection rating for the failure mode.

The first of these activities relies very much upon the team contribution. Different individuals will bring different insights in terms of what can go wrong and what the effect will be. Painting this 'rich picture' of the product or process in terms of potential failures, effects and controls is an extremely beneficial quality planning activity.

The second activity results in the calculation of a **risk priority number (RPN)** which is simply the product of the severity, occurrence and detection ratings for the particular failure mode:

$$\text{RPN} = (\text{Severity}) \times (\text{Occurrence}) \times (\text{Detection})$$

The ratings given for severity, occurrence and detection are normally scored on a scale of 0–10 according to some descriptive and quantitative criteria as demonstrated in Table 7.1.

The failure risk priority is therefore an equally weighted combination of how severe the failure is in terms of its effect, how likely the failure is to occur and, if it does occur, how likely the process control is to detect the failure. A severe type of failure which rarely occurs and is easy to detect would therefore have a lower priority than would be the case if the effect of the failure alone was considered.

An example of a process FMEA developed for a batch manufacturing company is shown in Table 7.2.

7.2.3 IDENTIFYING CRITICAL CHARACTERISTICS

The output from the first six stages of the FMEA is the value of the RPN. The normal evaluation criteria for the RPN is:

$$\text{RPN} > 90 \text{ requires corrective action}$$

Table 7.1 Typical criteria for severity, occurrence and detection

	Ranking	Possible failure rates
Severity of effect		
Minor: Unreasonable to expect that the minor nature of this failure would cause any real effect on the system performance, or on a subsequent process or assembly operation. Customer will probably not even notice the failure.	1	
Low: Low severity ranking due to nature of failure causing only a slight customer annoyance. Customer will probably only notice a slight deterioration of the vehicle system performance, or a slight inconvenience with a subsequent process or assembly operation, i.e. minor rework action.	2 3	
Moderate: Moderate ranking because failure causes some customer dissatisfaction. Customer is made uncomfortable or is annoyed by the failure. Customer will notice some subsystem performance derioration. May cause the use of unscheduled reworks/repairs and/or damage to equipment.	4 5 6	
High: High degree of customer dissatisfaction due to the nature of the failure or an inoperable convenience subsystem. Does not involve safety or non-compliance to government regulations. May cause serious disruption to subsequent processing or assembly operations, require major reworks, and/or endanger machine or assembly operator.	7 8	
Very high: Very high severity ranking when a potential failure mode affects safe operation and/or involves non-compliance with government regulations	9 10	
Probability of failure		
Remote: Failure is unlikely. No failure, ever associated with almost identical processes. $C_{pk} \geq 1.67$	1	< 1 in 10^6 ~±5s
Very low: Process is in statistical control. Capability shows a $C_{pk} \geq 1.33$. Only isolated failures associated with almost identical processes.	2	1 in 20000 ~±4s
Low: Process is in statistical control. Capability shows a $C_{pk} \geq 1.00$. Isolated failures associated with similar processes.	3	1 in 4000 ~±3.5s
Moderate: Generally associated with processes similar to previous processes which have experienced occasional failures, but not in major proportions. Process is in statistical control with a $C_{pk} \leq 1.00$.	4 5 6	1 in 1000 ~±3s 1 in 400 1 in 80
High: Generally associated with processes similar to	7	1 in 40

previous processes that have often failed. Process is not in statistical control.	8	1 in 20
Very high: Failure is almost inevitable.	9	1 in 8
	10	1 in 2

Where: C_{pk} is the process capability index.

Detection of failures

Very high: Controls will almost certainly detect the existence of a defect. (Process automatically detects failure.)	1
	2
High: Controls have a good chance of detecting the existence of a defect.	3
	4
Moderate: Controls may detect the existence of a defect.	5
	6
Low: Controls have a poor chance of detecting the existence of a defect.	7
	8
Very low: Controls probably will not detect the existence of a defect.	9
Absolute certainty of non-detection: Controls will not or cannot detect the existence of a defect.	10

For a process FMEA the requirement for corrective action generally implies an improvement in the likelihood of occurrence or the ability to detect. In most practical situations the severity of the occurrence is determined by the effect upon the customer and is difficult to improve (reduce) in value. An important factor in the likelihood of occurrence and the ability to detect faults is the extent to which the process is controlled. Where an RPN has a value greater than 90, this normally denotes an aspect of the process requiring improved control and in many applications this implies SPC.

The format for identifying the critical characteristics of the process is normally the **control plan**. The control plan tabulates those characteristics identified through the FMEA which have a high value of RPN. For each of these characteristics the control plan should define:

- method of determination (testing);
- point of testing;
- frequency of testing;
- analysis method;
- specification tolerances;
- reaction to 'out of specification' condition.

In consultation then with the customer the critical characteristics are then

Table 7.2 Example of a process FMEA

Process	Purpose	Potential fail mode	Effects of failure	Causes of failure	Controls	Existing S O D R / E C E P / V C T N	Recommended actions	Result S O D R / E C E P / V C T N
Premix	● Add ingredients used ● Basic composition	● Incorrect ingredients used	● Variable tensible strength ● Variable compactability SMALL EFFECT ● Low durability of the clay	● Human error or incorrect labelling ● Incorrect raw materials	● Process mill ticket ● Acceptance criteria at G.I.	5 5 7 175 8 2 2 32	● Implement mill ticket as production control document ● No action planned	5 5 3 75
		● Incorrect raw material quantities	● Incorrect dry strength ● 'Shake out' problems	● Incorrect weighing (human or equipment)	● Methylene blue tests and X and R charts (required)	5 4 7 140	● Implement control charts	5 4 2 20
		● Cross contamination of raw materials	● Very small effects as per incorrect ingredients above	● Overstocking of quantities	● Appropriate stockpiling	2 5 3 30	● No action planned	
Milling	● Particle size ● Homogenous mixture	● Grain size too large	● Conveying problems ● Slow development	● Incorrect adjustments (fan too high or whizzer too low)	● SPC X and R charts (moving averages) for grain size	7 5 7 245	● Implement control charts	7 5 2 70
		● Contamination of colour ● Grain too small	● Potential minor contamination ● Dusty ● Excess extraction losses	● Mechanical breakdown ● Ineffective flushing ● Incorrect fan settings	● Visual inspections product ● Grain size control charts for x and R	3 5 3 45 6 5 7 210	● No action planned ● Implement control charts	 6 5 2 60
Blending	● Composition	● Wrong composition	● Incorrect dry strength ● 'Shake out' problems ● Moulding problems	● Incorrect settings ● Mechanical problems	● Specifications of recipes & settings ● Methylene blue tests and x and R charts	5 4 7 140	● Implement mill ticket ● Implement control charts	5 4 2 40 7 4 3 84
Packing	● Quantities	● Incorrect weights	● Material shortages	● Packing errors (human or equipment)	● Calibration of scales & weighbridges	2 5 3 30	● No action planned	

QUALITY CONTROL PLAN

PROCESS: Milling and Blending Powdered Clay Products

CUSTOMER:

PRODUCT: Duraclay 7

ISSUE NO: 4

SHEET: 1 of 1

CHARACTERISTIC	LEVEL*	TEST METHOD	SAMPLING POINT	FREQUENCY	ANALYSIS METHOD	SPECIFICATION TOLERANCES	REACTION TO OUT OF CONTROL CONDITION
PARTICLE SIZE	1	% RETAINED ON 150 MICRON SLEEVE	MILL	COMPOSITE OF SAMPLES FROM PRODUCTION RUN	X AND R CHARTS	1.0 - 6.0% RETAINED	ADJUST MILL SETTING/INVESTIGATE MECHANICAL CONDITION
METHYLENE BLUE	2	TITRATION	MILL	COMPOSITE OF SAMPLES FROM PRODUCTION RUN	RUN CHART AND LAB RECORD SHEET	$42.5 \pm 3.5 cm^3$	INVESTIGATE RAW MATERIAL QUANTITIES AND PROPERTIES AND ADVISE
WET TENSILE	2	WET TENSILE TEST EQUIPMENT	MILL	COMPOSITE OF SAMPLES FROM PRODUCTION RUN	LAB RECORD SHEETS	0.27 to 0.35 N/Cm^2	INVESTIGATE RAW MATERIAL QUANTITIES AND PROPERTIES AND ADVISE
MOISTURE	2	WEIGHT LOSS DURING CONTROLLED DRYING	MILL	COMPOSITE OF SAMPLES FROM PRODUCTION RUN	LAB RECORD SHEETS	15% MAX	INVESTIGATE RAW MATERIAL AND ADVISE

1. DENOTES SIGNIFICANT CHARACTERISTIC
2. DENOTES IMPORTANT CHARACTERISTIC

ISSUED:................................ DATE:............................ CUSTOMER APPROVAL.................................. DATE:...................

Figure 7.4 Quality control plan.

defined by considering the data from both the FMEA and the control plan. An example of a control plan for the process FMEA shown above in Table 7.2 is shown in Figure 7.4.

Some sectors of industry have modified this process of identifying critical characteristics using firstly the FMEA and then a control plan to suit industry-specific requirements. For example, the food industry uses a failure mode analysis technique known as 'hazard analysis and critical control points' (HACCP) to identify, monitor and control areas of the food production process which may be critical in terms of the likelihood of problems or contamination.

7.2.4 SUMMARY

- Failure mode effects and analysis provides a pinpointing technique to help identify the process parameters which require improved control.
- The process FMEA uses teamwork to identify potential process failures and to rank these in terms of severity, occurrence and detection probabilities of failure.
- Failure modes with high risk priority numbers require control plans to plan and identify corrective actions.

7.3 Data collection and analysis

7.3.1 DATA TYPES AND APPLICATIONS

Having decided 'what' to measure and control from the identification of the critical characteristics, the next issue to consider is 'how' to measure the important parameters. The first criteria to consider is the degree of detail to which the characteristic is measured. Generally this involves the selection of either:

- **attribute data** in which the parameter is measured as either passing or failing (go/no-go) a pre-defined specification; or
- **variables data** in which the precise measurement is taken against a continuous scale (such as size, weight, time or other property).

Statistical process control is applied to both attribute and variables data and in general the decision as to which type of data to use depends upon the balance between the cost of sampling versus the cost of inspection as discussed in Chapter 6.

In order to effectively control the variation of a process, it is necessary to consider both:

- the **position**, which is generally measured in terms of the process average;

- the **spread**, which is generally measured in terms of range or standard deviation.

Controlling the variability of a process generally involves monitoring both of these aspects of accuracy and repeatability. A process can be deemed to be 'out of control' if:

- there is a change in process position;
- there is a change in process spread;
- there is a change in both position and spread.

The effects of changing position and spread is illustrated in Figure 7.5. The importance of monitoring both of these factors lies in the implication for corrective action. Generally a change in process position implies the need for the process to be reset whereas a change in spread implies the need for the disturbances to the process to be examined.

7.3.2 FREQUENCY DISTRIBUTION

The essence of SPC is to monitor the variation of the process and to be able to distinguish between:

- **Normal variation** – generally due to random combinations of minor factors and representing the usual variability of the process, materials or environment;

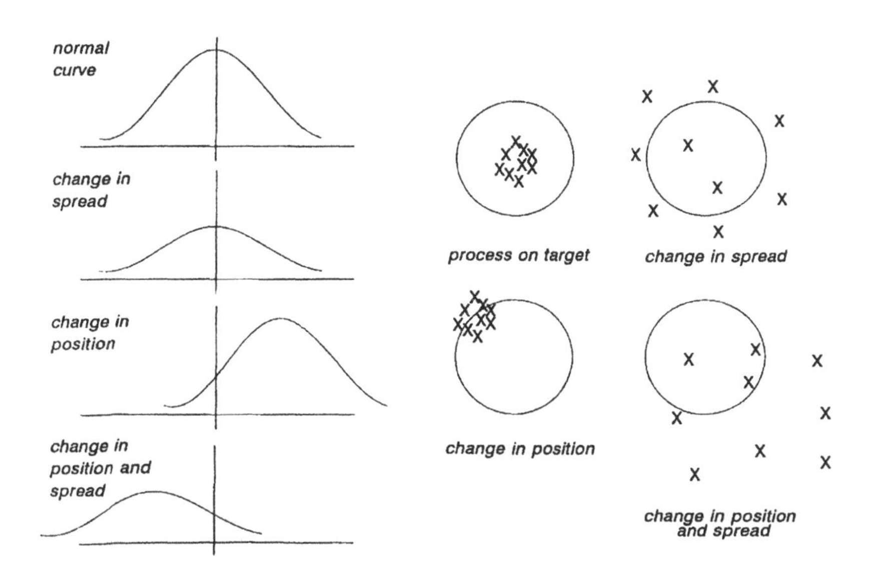

Figure 7.5 Process position and spread.

- **Special variation** – generally due to significant changes to the process operation or settings and representing unusual variability of the process, materials or environment.

The simple concept that, by monitoring the variation of the process, it is possible to distinguish between normal and special (assignable) variation is in practice rather more complex. At the outset of the collection of the data it is important to 'bound' the process and to define which factors and influences are to be considered a 'normal' part of the process variability. In reality it is often convenient to define the 'normal process' to include all those factors (such as temperature, humidity, material condition, etc) which cannot be readily controlled. These factors may, however, represent a significant cause of the overall variation of the process and therefore influence the capability to meet requirements.

Having defined the boundary of the process to be controlled, the critical issue is to model the way in which the process normally varies. The distribution most obviously used to describe processes subject only to random variation is the **normal (or Gaussian) distribution**. The main features of this distribution are that;

- it is symmetric about the mean value;
- it is representative of much of the variation in natural or industrial processes.

If data is collected from a process which is varying normally and plotted as a histogram, then the result would be as shown in Figure 7. 6. If the sample population is large then the distribution approximates to the classical normal curve.

For normally distributed data, the key measures of position and spread respectively are:

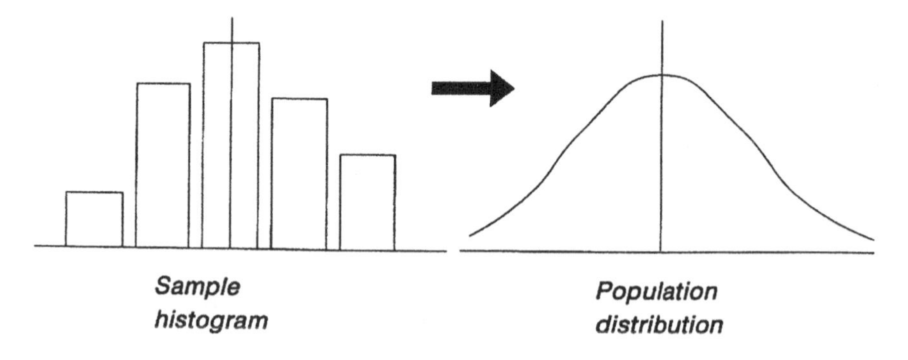

Figure 7.6 The normal frequency distribution.

- **mean value,** $\overline{X}$, where:

$$\overline{X} = \frac{\Sigma_{i=1}^{n} X_i}{n} = \frac{X_1 + X_2 + X_3 + \ldots + X_n}{n}$$

- **standard deviation,** σ, where:

$$\sigma = \sqrt{\frac{\Sigma_{i=1}^{n} (\overline{X} - X_i)^2}{n-1}}$$

The significance of the standard deviation is that it can be used to indicate the probability that a process value would 'normally' fall within the range measured in terms of σ. If the standard deviations are drawn onto the normal distribution then the probability of a value occurring within the range is shown in Figure 7.7.

An alternative interpretation of the limits measured in terms of the standard deviation is:

- approximately 1 in 40 of the normal population lie outside the limits of either $+2\sigma$ or -2σ (the actual value is 1.96 standard deviations);
- approximately 1 in 1000 of the normal population lie outside the limits of either $+3\sigma$ or -3σ (the actual value is 3.09 standard deviations).

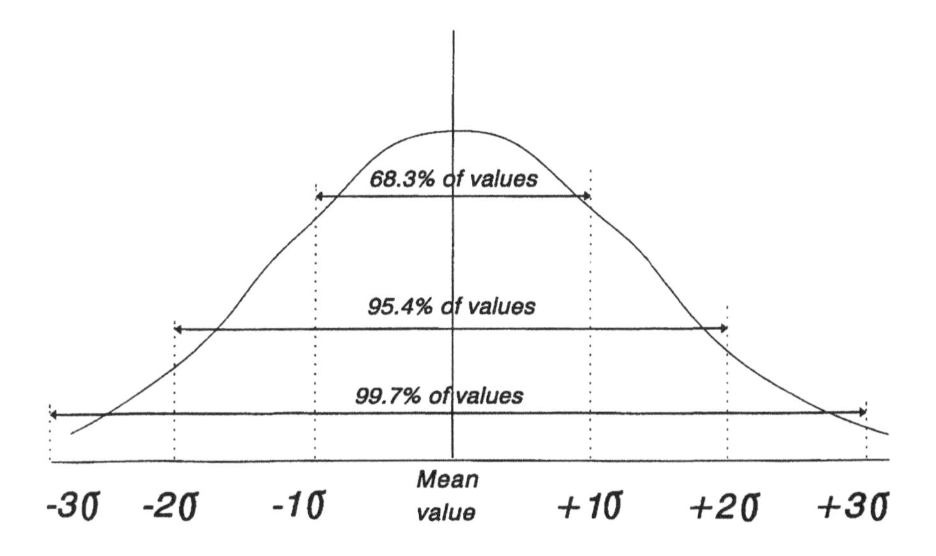

Figure 7.7 Standard deviation of the normal distribution.

It is the practical implications of these probabilities that is the essence of SPC. For example, if a process is assumed to be operating normally, then if a value is found to be more than two standard deviations away from the average value, this acts as a warning because this has only a 1 in 40 chance of occurring naturally. Hence $\pm 2\sigma$ represents the **warning limits.**

Similarly, if a value is found to be more than three standard deviations away from the average value, this indicates some process corrective action is required because this value has only a 1 in 1000 chance of occurring naturally. Hence $\pm 3\sigma$ represents the **action limits.**

7.3.3 EVALUATING DATA

The principle of modelling processes using the normal distribution and then identifying 'out of control' conditions using the probability implication of two and three standard deviations is both very simple and very powerful. The underlying assumption, however, in applying this SPC approach is that the process data is distributed normally. An important prerequisite therefore to implementing SPC is establishing whether the process is normal. The most commonly used techniques for testing whether data are normally distributed are:

- skewness and kurtosis;
- probability plots;
- chi-squared tests.

The **skewness** of the data is a measure of the symmetry of the process data. The most common cause of a process not being symmetric is the presence of some form of natural cut-off. For example, monitoring the thickness of a plastic coating which on average is, say, 0.5 mm but may increase to, say, 1.5 mm is distorted by the fact that the thickness of the coating may not fall below zero. This natural data cut-off at zero thickness would distort the probability of values falling either above or below the mean value. Skewness is the measure of distortion and is given by:

$$\text{Skewness} = \frac{\Sigma_{i=1}^{c} f_i (X_i - \overline{X})^3 / n}{\sigma^3}$$

where f_i is the frequency of the ith cell of data and c is the number of cells. A cell is a group of data values which fall within limits within the total range of the data.

The value of skewness can then be used to evaluate the symmetry as follows:

Skewness	=	0	data is symmetric
Skewness	>	0	data is skewed to the right
Skewness	<	0	data is skewed to the left

Where the absolute value of skewness approaches unity this indicates a strongly skewed data distribution.

The **kurtosis** of the data is a dimensionless measure of the extent to which the distribution peaks. The kurtosis is given by:

$$Kurtosis = \frac{\Sigma_{i=1}^{c} f_i (X_i - \overline{X})^4 / n}{\sigma^4}$$

The normal distribution has a kurtosis value of 3. If the data show a value of kurtosis less than 3, then the distribution is flatter than normal. If the value of kurtosis is greater than 3, then the distribution is more peaked than normal. The measures of skewness and kurtosis are illustrated in Figure 7.8 together with an example of calculating the values.

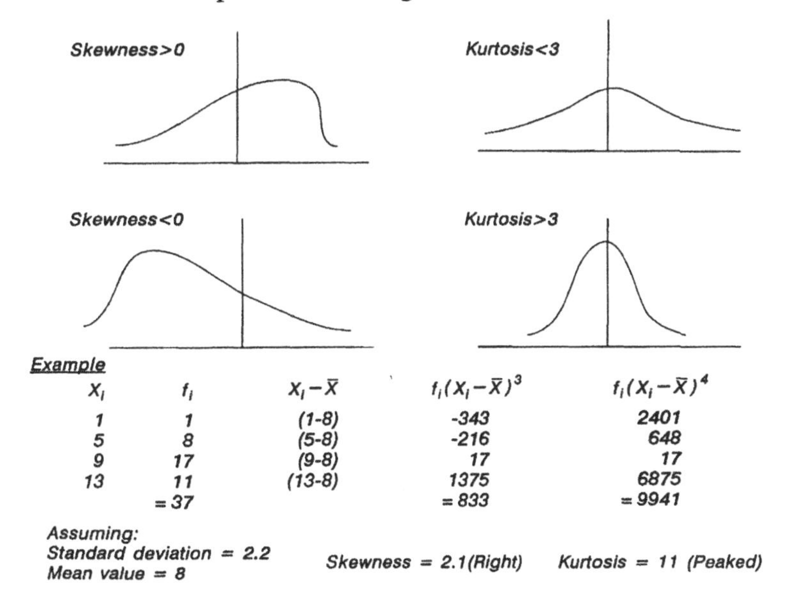

Figure 7.8 The skewness and kurtosis of the data distribution.

Probability plots are a graphical technique used to assess the extent to which a set of data is distributed normally. Probability plots use a special type of Chartwell graph paper which plots data values against percentage of values (plotting position). To generate a probability plot, firstly the data should be ordered from smallest value to largest value and the plotting position is then calculated from:

$$\text{Plotting position} = \frac{100(i - 0.5)}{n}$$

where i is the ordered position of the data and n the total sample size. If the data is distributed normally, then the probability plot will be a straight line as illustrated in Figure 7.9. In practice some judgement is generally required to evaluate plots and as with Weibull plots (Chapter 9) the key distinction is the extent to which the data evenly straddles the best-fit straight line rather than exhibiting a defined curvature.

The **chi-squared** goodness-of-fit test is used to test the hypothesis that the frequency of each cell is the normally expected value. To carry out the chi-squared test, each of the cells should contain at least five data values and the frequency of values should be compared with the expected frequency from the normal distribution tables (Appendix A).

In many industrial situations, the data are collected in groups (samples) for which the average value can be calculated. The importance of this is that even where the individual process data values are not distributed normally, the distribution of sample means (averages) is usually normal. This property of mean values being normally distributed is given by the **central limit theorem**.

Finally where the process is monitored using groups (samples) of data, then a simple measure of variability is **sample range** where:

$$\text{Range} = \text{Largest value within the sample} - \text{Smallest value}$$

The average value of the range is therefore an easily determined measure of process spread. The value of the range does, however, depend upon the sample size (generally increasing with sample size) and is therefore not such a consistent measure of data variability as standard deviation. Due to

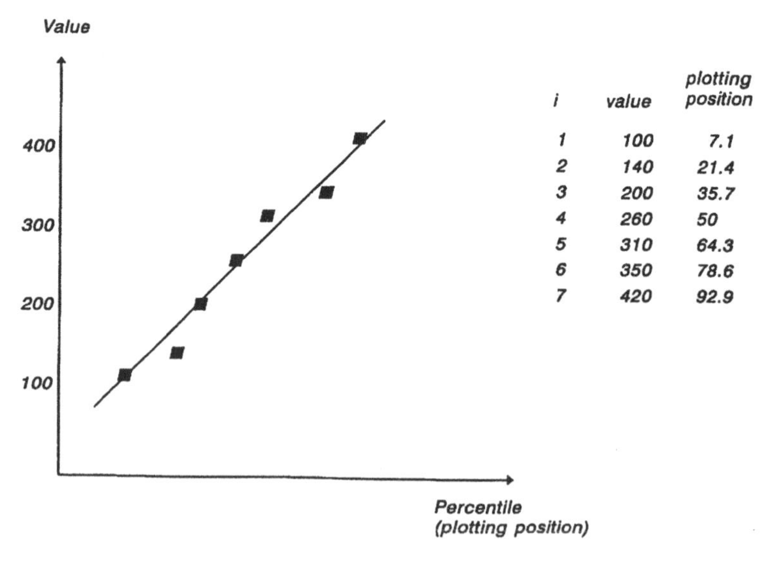

i	*value*	*plotting position*
1	100	7.1
2	140	21.4
3	200	35.7
4	260	50
5	310	64.3
6	350	78.6
7	420	92.9

Figure 7.9 Probability plot.

its ease of computation, the average value of the range is often used to approximate the value of the standard deviation of the sample means using:

$$\sigma_{\overline{X}} = \frac{\overline{W}_X}{d_n \sqrt{n}}$$

where: $\sigma_{\overline{X}}$ = the standard deviation of the sample means;
$\overline{W}_X$ = the average of the sample ranges;
d_n = a constant depending upon the sample size, n.

Tables showing the constant d_n are available to simplify the calculation of the standard deviation in industry. Both the constant d_n and also a second constant A_n where:

$$A_n = \frac{3}{d_n \sqrt{n}}$$

are shown in Table 7.3. Hence, the control limits are: $\overline{X} \pm A_n \overline{W}_X$.

Table 7.3 Values of constants

Sample size n	d_n	A_n
2	1.13	1.88
3	1.69	1.02
4	2.06	0.73
5	2.33	0.58
6	2.55	0.48
7	2.70	0.42
8	2.87	0.37
9	2.94	0.34
10	3.06	0.31

So, for example, consider the set of data shown in Table 7.4. The control limits for the process can be determined from simple calculations of the sample averages and the sample ranges as illustrated. The control limits for this data would therefore be:

$$\text{UCL} = \overline{X} + A_n \overline{W}_X = 20 + 1.02 \times 3 = 23.06$$
$$\text{Centreline} = \overline{X} = 20$$
$$\text{LCL} = \overline{X} - A_n \overline{W}_X = 20 - 1.02 \times 3 = 16.94$$

Table 7.4 The use of control limit constants

Sample number	Reading 1	Reading 2	Reading 3	Sample mean	Sample range
1	18	20	22	20	4
2	17	25	21	21	8
3	19	19	19	19	0
4	19	18	17	18	2
5	20	22	24	22	4
6	20	20	20	20	0
			Overall mean	20	
			Mean of ranges		3

7.3.4 SUMMARY

- The essence of statistical process control is distinguishing between normal (random) variation and special (assignable) variation.
- The normal distribution can be used to represent the variation present in many industrial processes and confidence levels can be set at $\pm 2\sigma$ (warning limits) and at $\pm$ 3σ (action limits).
- Process data can be tested for normality using skewness, kurtosis, probability plots or chi-squared tests.

7.4 Process capability and indices

7.4.1 THE ROLE OF PROCESS CAPABILITY IN SPC

Having identified 'what' to measure through FMEA and 'how' the process is varying through the examination of the data distribution, the next point to consider is whether the variability of the process is acceptable. By comparing the natural variability of the process with the specification, a measure of the ongoing capability to conform to requirements is obtained.

This capability of the process is an important factor in the implementation of SPC in that it provides:

- a measure of process improvement;
- a focus for pinpointing process improvements;
- criteria for process acceptance.

Increasingly the capability of the process is used in industry to evaluate both new processes and new suppliers. If the process capability is either very good or very poor, then in either case it is unnecessary to proceed to the next stage of SPC implementation, namely the construction of control charts. Processes with very poor capability require corrective actions prior to the implementation of routine monitoring using control charts. Conversely a process with very good process capability in most practical cases does not necessarily justify the time and effort involved in setting up and operating control charts.

Increasingly in certain sectors of industry, most notably the motor and electronics industries, the proof of an acceptable level of process capability is a basic requirement for suppliers. The reason for this from the customer viewpoint is that suppliers whose process variability is small compared to the requirements specification will statistically be able to routinely deliver conforming materials time after time. The assurance of quality associated with good process capability has made this a widespread vendor assessment and selection criterion.

7.4.2 PROCESS CAPABILITY AND CAPABILITY INDEX

In simple terms the capability of the process is the ratio of the process specification or tolerance to the natural variability of the process. The specification or tolerance designated for the output from the process represents a 'window' through which the process variability must pass. For a process which is varying normally 99.7% of the data will fall within the range ±3 standard deviations. This spread of the process is usually termed the **effective range**. The process capability is therefore given by:

$$\text{Process capability } (C_P) = \frac{\text{Total specification tolerance}}{\text{Total effective range}}$$

$$\text{hence } C_P = \frac{T_U - T_L}{6\sigma}$$

Where T_U and T_L are the upper and lower specification limits respectively. The process capability, C_P is then evaluated as follows:

$C_P = 1$ process is capable
$C_P << 1$ process not capable
$C_P >> 1$ process very capable.

The process capability can be improved by either reducing the variability of the process (the effective range) or by increasing the tolerance allowed. Capability is therefore an important design consideration and the implementation of SPC provides many manufacturing organizations with an improved understanding of the relationship between design and manufacture. Service companies can use the concept of capability to improve the integration of process operation and customer satisfaction. A transport company, for example, can provide a more reliable service if the variability in journey time is incorporated into the timetabled schedules.

A further refinement in assessing the capability of a process is to also consider the extent to which the variability is centred within the required specification range. This measure of centring is termed the **process capability index** and is given by:

$$\text{Process capability index } (C_{Pk}) = \frac{\text{Average value} - \text{Nearest tolerance}}{\text{Half the effective range}}$$

$$\text{Hence } C_{Pk} = \frac{\bar{X} - T_L}{3\sigma}, = \frac{T_U - \bar{X}}{3\sigma}, \text{ whichever is smaller.}$$

If the process is correctly centred then the process capability and the capability index are equal. In general, however, the value of C_{Pk} is usually smaller than the value of C_P due to the process not operating in the centre of the tolerance band.

The relationship between process capability and capability index is illustrated in Figure 7.10.

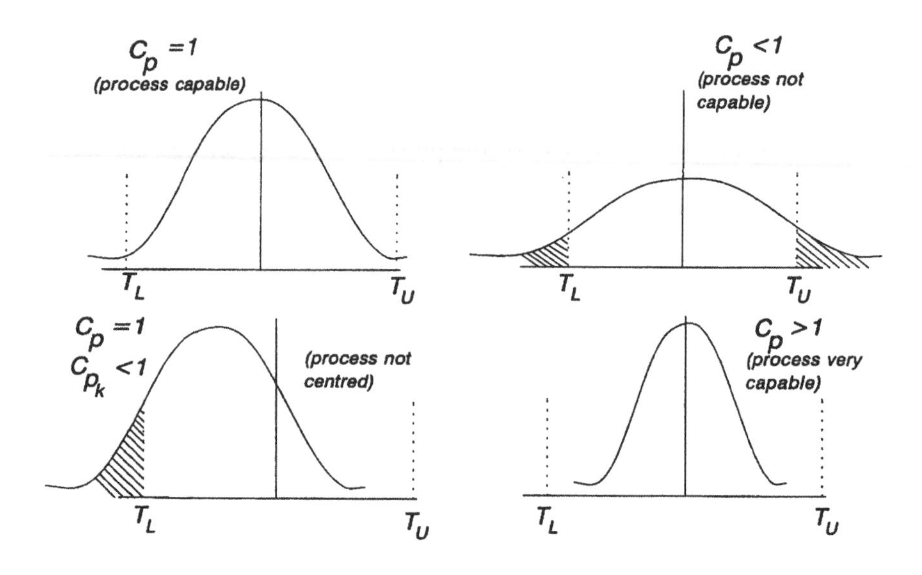

Figure 7.10 Process capability and capability index.

7.4.3 INTERPRETATION OF PROCESS CAPABILITY INDICES

The capability of processes and suppliers is seen by many organizations as fundamental to the quality development process. The Motorola quality improvement programme, for example, is called the 'six sigma' policy in which the company strives for a six sigma level of error-free performance in order to promote customer satisfaction.

The concept of process capability is seen as an important consideration in:

- the selection of raw material suppliers;
- the selection and acceptance of new processes;
- the design of product tolerances.

The use of C_p as an improvement measure is also an important part of the stage of quality development associated with creating an improvement orientation. Capability studies are commonly undertaken as part of the activities of quality improvement teams.

Using process capability as an input to the design of products supports the prevention orientation and enables more robust designs to be produced which take into account the inherent variability of manufacturing processes.

7.4.4 SUMMARY

- Establishing the capability of the process is an important initial stage in the implementation of SPC.
- Process capability is the ratio of the allowed tolerance to the effective range and the capability index is a measure of how well the process is centred within the tolerance range.
- Process capability is an important measure of quality development which can be applied to suppliers and to product design in addition to the evaluation of manufacturing processes.

7.5 Process control charts

7.5.1 MEANS AND RANGE CHARTS

The ongoing monitoring and control of the critical characteristics of a process is provided using **control charts**. Control charts are a graphical approach to monitoring the behaviour of the process by comparing the ongoing variation with warning and action limits derived from the normal distribution. The most basic form of control chart is the 'means chart' for which the process is sampled and the average value of the sample data is plotted sequentially. Limits are drawn onto the means chart at:

- $\pm$ 1.96 standard deviations, which corresponds to the warning limits;
- $\pm$ 3.09 standard deviations, which corresponds to the action limits.

The value of 1.96 and 3.09 standard deviations correspond to probabilities of 1 in 40 and 1 in 1000 precisely. In the USA it is more common to set action limits at 3 standard deviations to simplify the calculation of limits although the probability associated with this level of deviation is slightly less than 1 in 1000. The control limits on a means chart are therefore given by:

- Upper action limit $\overline{\overline{X}} + 3.09\sigma_{\overline{X}} = \overline{\overline{X}} + \dfrac{3.09\overline{W}_X}{d_n\sqrt{n}}$
- Upper warning limit $= \overline{\overline{X}} + 1.96\sigma_{\overline{X}} = \overline{\overline{X}} + \dfrac{1.96\overline{W}_X}{d_n\sqrt{n}}$
- Centreline $= \overline{\overline{X}}$
- Lower warning limit $= \overline{\overline{X}} - 1.96\sigma_{\overline{X}} = \overline{\overline{X}} - \dfrac{1.96\overline{W}_X}{d_n\sqrt{n}}$
- Lower action limit $= \overline{\overline{X}} - 3.09\sigma_{\overline{X}} = \overline{\overline{X}} - \dfrac{3.09\overline{W}_X}{d_n\sqrt{n}}$

The stages in the preparation of control charts for means are as follows:

- **Stage 1**. Conduct a pilot study with the process in a known state of control and take (approximately) 25 samples.
- **Stage 2**. For each sample determine the average, X and the range W_X.
- **Stage 3**. Calculate the overall average X and the average of the range value W_X.
- **Stage 4**. Calculate the process capability to ensure the process is viable.
- **Stage 5**. Construct the means control charts using the limits described above.

As described above in section 7.3.1, the control of most processes requires the monitoring of both position and spread. Position is measured using the means chart and spread is measured using the 'range chart'. The control limits for the range chart are given by the following:

- Upper action limit $= D_4 \overline{W}_X$
- Upper warning limit $= D_3 \overline{W}_X$
- Lower warning limit $= D_2 \overline{W}_X$
- Lower action limit $= D_1 \overline{W}_X$

where D_1, D_2, D_3 and D_4 are constants.

In general, only the upper action and warning limits are used for range charts as the lower limits imply that the variation of the process has reduced and generally this would only be investigated to examine whether an inspection error had occurred or some unexpected improvement had taken place. Typical means and range charts are shown in Figure 7.11.

Means and range charts are the most commonly used form of variables control chart and this is due to the central limit theorem which states that such data is typically normally distributed and is therefore appropriate to

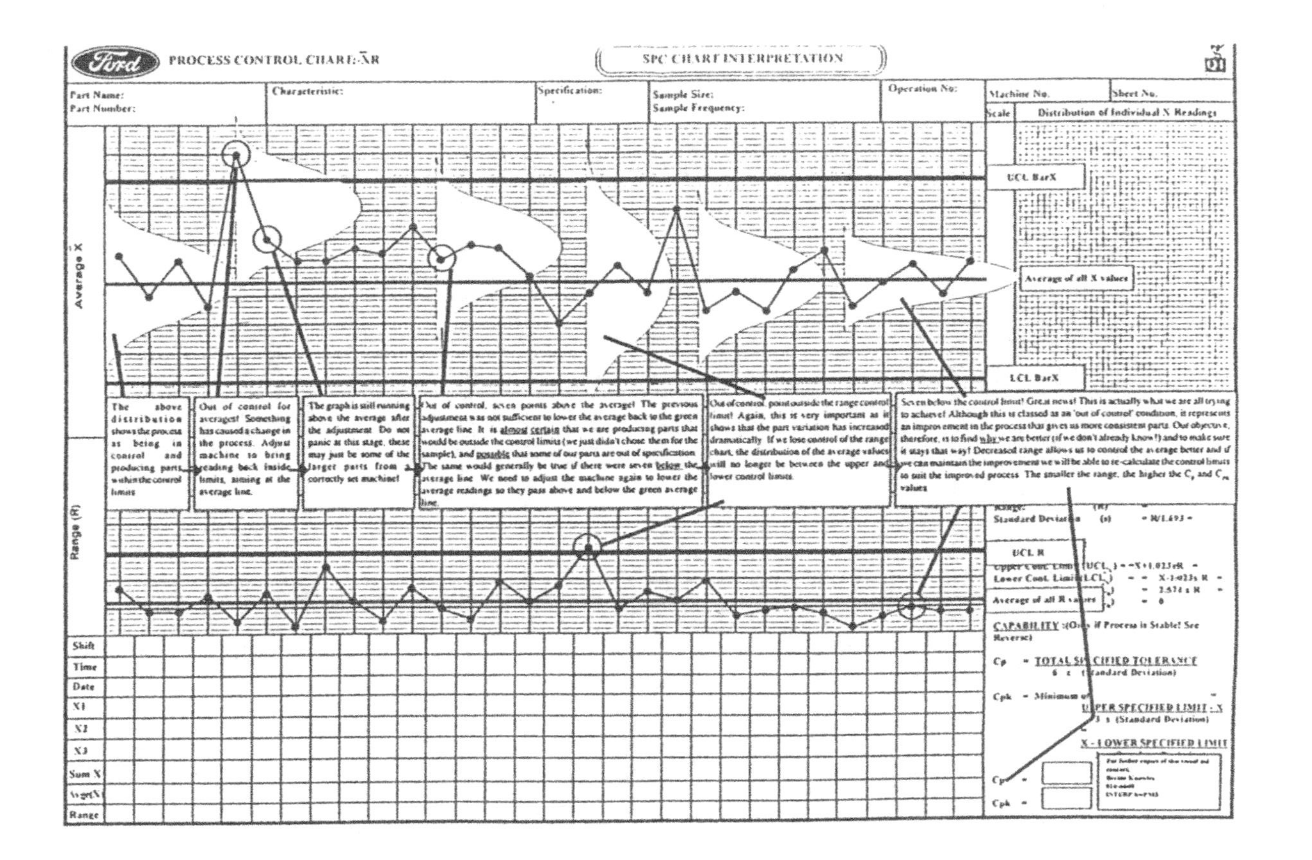

Figure 7.11 Means and range chart. (Reproduced by kind permission of Ford Motor Company Ltd, Halewood, UK.)

this form of control. There are, however, other forms of variables control charts most notably:

- individual control charts;
- moving average and moving range charts.

The individual control chart, as the name suggests, is implemented by plotting the individual values rather than the average of a sample. Individual data is used in situations where it is not possible (either technically or financially) to obtain more than one data value per item. Such charts are generally more difficult to interpret and show much greater variability than an equivalent means chart. In addition, with individual data it is not possible to calculate range values and in practice usually a moving range using two data points is generally employed. Individual control charts are therefore generally less useful and only used where other forms are not available.

Moving average and moving range charts are typically used in the batch process industries in which the testing of the product may take place over a period of time and a considerable quantity of product would be produced before sufficient data becomes available to plot a sample average. By using moving average or moving range values, each individual data value can be incorporated into the moving average or moving range and therefore can be plotted onto the control chart.

Figure 7.12 illustrates the way in which the moving average and moving range approach produces more data points than the corresponding means or range charts for the equivalent sample size. This approach is therefore most applicable where a considerable time elapses between data samples and also has a 'smoothing' effect when compared to plotting an equivalent individual chart.

7.5.2 CONTROL CHARTS FOR ATTRIBUTE DATA

In certain practical applications it is neither physically possible nor economically feasible to obtain precise measurement of the critical characteristic (variables data) and instead attribute data is used. Where defective product or service is identified (for example, a damaged component on the production line or a customer service complaint), it often does not make any practical sense to attempt to precisely measure the level of defectiveness (the extent of the damage or the dissatisfaction of the customer). In such instances, the statistical control of the process is undertaken using attribute data and generally in such applications only a single chart is used as range data is not available.

The two most commonly used forms of the attribute data control chart are:

- the proportion defective chart (p-chart);

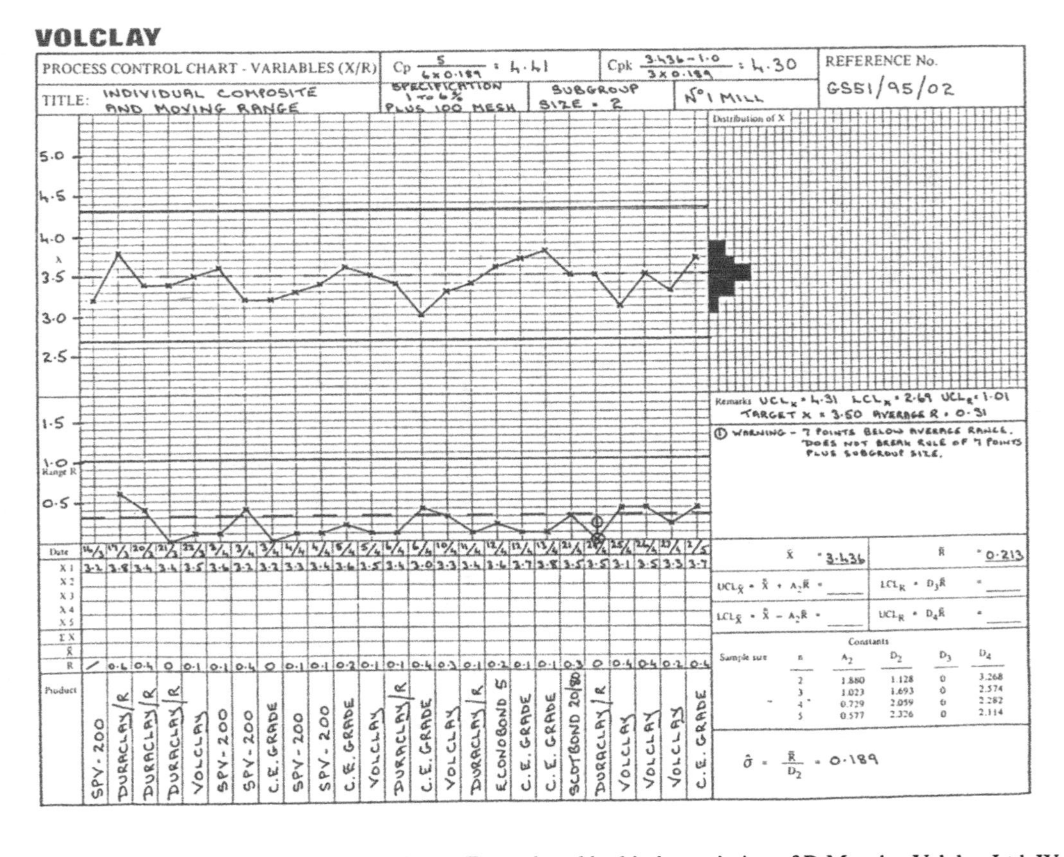

Figure 7.12 Moving average and moving range charts. (Reproduced by kind permission of P. Murphy, Volclay Ltd, Wallasey, UK.)

- the number of defects chart (c-chart).

The **proportion defective control chart** is applicable where the number of defects found can be expressed as a proportion of the sample size. This approach has the advantage of being applicable in situations where the sample size is not necessarily constant (for example, in batch production industries where the batch size may vary to suit specific customer order quantities).

The proportion defective or p-chart is constructed by conducting a pilot study to determine the average proportion defect, p, over several samples. Therefore:

$$\text{Mean proportion defective } (\overline{p}) = \frac{\text{Total number of defects within (say) } 20-30 \text{ samples}}{\text{Total number inspected}}$$

This pilot study to determine $\overline{p}$ should of course be conducted with the process in a known state of control. The standard deviation for this type of situation where the probabilities are concerned with the proportion of a group or sample is given by the binomial distribution and can be shown to be:

$$\text{Standard deviation of proportions, } \sigma_p = \sqrt{\frac{\overline{p}(1-\overline{p})}{n}}$$

which can be readily calculated from the average proportion defective, p, and the sample size, n. The control limits plotted on a proportion defective chart are normally only those associated with action limits and these are therefore given by:

- Upper action limit $= \overline{p} + 3.09\sqrt{\frac{\overline{p}(1-\overline{p})}{n}}$
- Centreline $= \overline{p}$
- Lower action limit $= \overline{p} - 3.09\sqrt{\frac{\overline{p}(1-\overline{p})}{n}}$

The most significant feature of this type of control chart is that the control limits are a function of the sample size and therefore (if the sample size is varying) need to be recalculated for each data value plotted onto the chart, as illustrated in Figure 7.13. To simplify this problem the average sample size is sometimes used to calculate fixed control action limits on the p-chart.

As with the range charts, the out of control points on a p-chart generally only apply to the upper action limit. Values below the lower action limit imply 'too few' defectives and would only be investigated to determine whether the process had improved or whether a monitoring error had occurred.

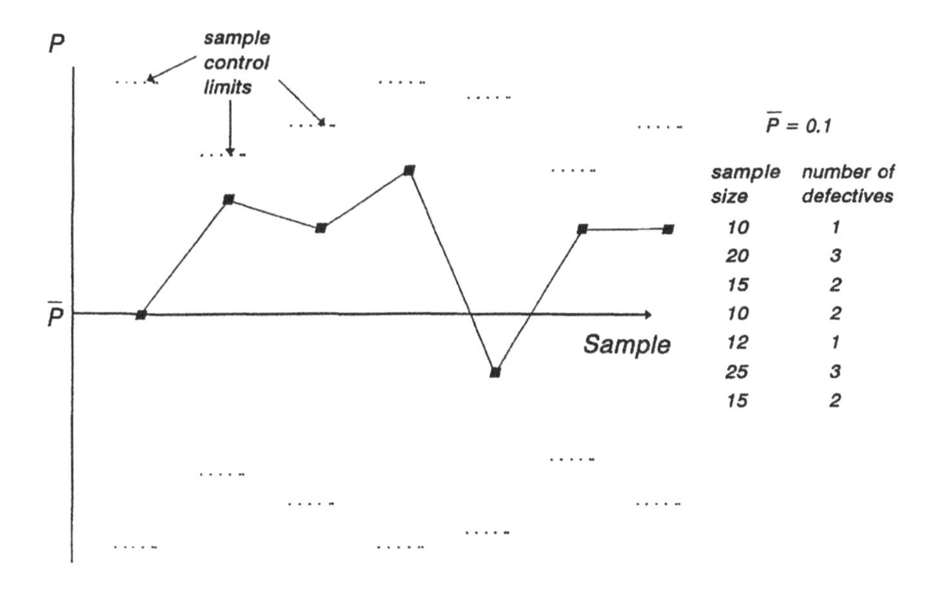

Figure 7.13 The proportion defective chart.

The **number of defects chart (c-chart)** is applicable where attribute data is to be used but where the number of defects found cannot be expressed as a proportion of the sample which contains other, non-defective components. For example, the number of minor scratch marks in the paint finish of a new car can be counted but cannot be expressed as a proportion of the number on 'non-scratches' on the paintwork. The c-chart is therefore commonly used where an item may have more than one defect (for example, in a length of cloth) without rendering the item defective. In situations such as this where it is possible to measure and observe the occurrence of defects but impossible to measure the number of 'non-defects', then the appropriate probability model is the Poisson distribution. For discrete events (such as a defect) which occur in some continuous interval of space or time, then the Poisson distribution gives a standard deviation based upon an average number of defects per unit (of length or time, say) as follows:

$$\text{Standard deviation of defects } (\sigma_c) = \sqrt{\bar{c}}$$

Where $\bar{c}$ is the average number of defects per unit and is obtained from a pilot study conducted to examine several (typically 20 to 30) units. The control limits for the c-chart are therefore given by:

- Upper action limits $= \bar{c} + 3.09\sqrt{\bar{c}}$
- Centreline $= \bar{c}$
- Lower action limits $= \bar{c} - 3.09\sqrt{\bar{c}}$

The normal industrial practice is to use the symmetric action limits given above despite the fact that the Poisson distribution is not symmetrically distributed. These limits are chosen in practice to make the approach to plotting c-charts consistent with other forms of statistical process control charts. In the construction and application of a c-chart as shown in Figure 7.14 it is important to note that the sample size (unit of length, time, etc.) has to remain constant.

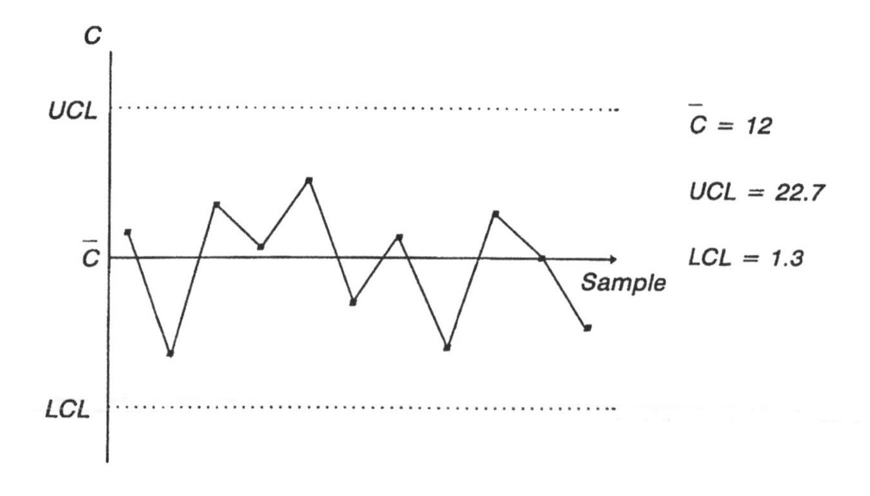

Figure 7.14 The number of defects chart.

7.5.3 THE INTERPRETATION OF CONTROL CHARTS

Process control charts are a graphical technique for the monitoring and control of process variability. By their visual nature they assist process operators, supervisors and managers to gain a greater understanding of the behaviour of the process and hence provide better control. For many processes the important decision is whether to take some form of corrective action or whether to leave the process alone. By using the statistical probability associated with a process value being a certain distance away from the normally expected average value then this type of decision becomes very much more assured.

The interpretation of process control charts basically involves the identification of statistically significant data values or patterns of data. The major statistically significant events on a control chart are as follows.

- Values **outside control limits** indicate the process has been subjected to a non-random, assignable influence. Values falling outside the warning limit only have a 1 in 40 chance of occurring naturally and would generally initiate some form of process investigation. Process values falling outside the action limits have only a 1 in 1000 chance of occurring normally and are therefore almost certainly an indication that the process is out of control. The specific action to be taken would of course depend upon the nature of the process and many organizations define in some form of quality system procedure (Chapter 2) what the prescribed corrective action should be. In general, when the process data falls outside the action limits the process is halted or the output quarantined (depending upon the tolerance limits) until the cause of the special variation is identified. A process can go out of control by incurring a change in mean value, a change in the process spread or a combination of both. The control limits therefore need to be monitored on both the means and range charts.

- Values may exhibit **trends** in which a run of process data presents a non-random pattern. The most widely used guideline for examining trends on control charts is the 'run of seven' rule. Given that each data value has a 50:50 chance of being above or below the process mean value then the probability of having seven consecutive points above or below the mean 0.5^7 or 1 in 128, which again is a probability unlikely to have occurred normally. Similarly the normal probability of a process value being greater (or smaller) than the previous value is 50:50 and therefore the chances of having seven consecutive points in a line either increasing or decreasing is again 1 in 128 and hence unlikely to represent normal variation. With such trend interpretation it is also possible to consider warning limits where five consecutive points are arranged.
- **cyclical trends** are where the data plotted on the control chart exhibit a recurring pattern, often some form of sinusoidal variation. Such cycles in the process data also indicate some significant external influence on the process which is therefore deemed to be out of control. A process which remains within the control limits but exhibits cyclical trends is normally investigated to determine the cause of the pattern of variation.

Approximately two-thirds of all the values plotted onto the control chart should lie in the middle third of the chart. This is again derived from the normal distribution for which the probability of a value falling within ± 1 standard deviation is 68.3%.

In addition to these general guidelines for interpreting process control charts, in practice process operators and supervisors develop a visual 'norm' of how a particular chart should appear. The human ability for recognizing abnormal patterns is also an important attribute in the interpretation of process control charts.

The different guidelines for interpretation are illustrated on the control chart shown in Figure 7.15.

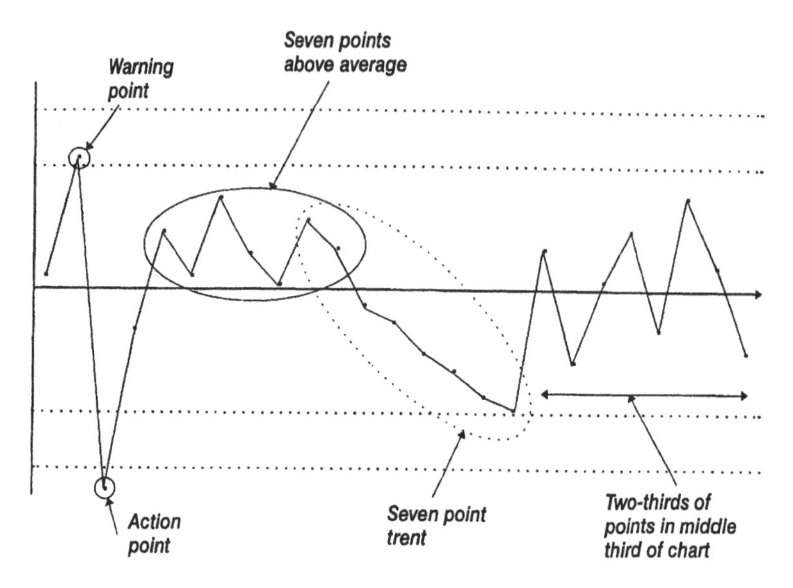

Figure 7.15 Guidelines for the interpretation of process control charts.

7.5.4 THE IMPLEMENTATION OF CONTROL CHARTS

Given that statistical process control is a simple yet powerful technique for monitoring and controlling process variability it is rather surprising that industrial implementation is not widespread. From an organizational perspective, the most common causes of ineffective SPC implementation programmes are:

- a lack of training and understanding for staff at all levels;
- a lack of a quality system base upon which to build the process controls;
- customer pressures to implement rather than internally recognized need;
- a lack of follow-up in terms of identifying and analysing the causes of variability which are often ignored once the process variation reduces.

Training, or rather the lack of it, is probably the most single critical issue. The people involved in the implementation of SPC not only need to be competent in the detailed techniques (FMEA, capability studies, control charting, etc.), but also understand the philosophy which links process variation, design specification and quality improvement.

The methods associated with SPC need to be formalized, preferably as part of the overall quality system documentation. These procedures need to describe not only the preparation of FMEAs, control plans, capability studies and control charts but also the prescribed investigation and corrective actions associated with a process being identified as out of control. SPC data should also form part of the management review of the overall process effectiveness.

In using SPC to improve the control of industrial processes it very soon becomes clear that a critical part of the process variability comes from the raw materials supplied. For this reason SPC very often has a cascading effect whereby implementing organizations begin to impose SPC requirements onto their suppliers. The danger in this methodology is that if the supplier implementation does not benefit from the same level of training/awareness or the same formalization into the quality system, the supplier commitment will be poor and the implementation superficial.

The implementation of the process control charts represents the final stage of the application of SPC (as described in section 7.1.3) although in most organizations the charts are the most significant ongoing manifestation of SPC. The parameters for the control charts which need to be initially established through some form of controlled pilot study are:

- mean values (either means, range, proportion, number or moving values as appropriate);
- control limits (normally warning and action limits set at ± 1.96 and ± 3.09 standard deviations respectively;
- process capability and capability index (normally expected to be greater than 1);
- data type (in terms of attribute or variables, averages or individual, fixed samples or moving samples);
- corrective actions to be taken in the event of the process reaching warning and action limits;
- responsibilities for completion, action and review of the control charts.

Such pilot studies would normally involve 20 to 30 samples being taken and the above parameters would be used to construct the initial control chart. Each control chart would typically contain 20–30 sets of process data and at the end of this control period it is usual to construct the next chart from the data generated from the previous chart (excluding any out-of-control points). This process of generating a rolling control chart from the data recorded on the previous chart is preferred in industry to using the original (pilot study) data as this approach more meaningfully reflects the evolution and improvement in the process. The original chart data established during the pilot study is often retained, however, as a reference benchmark.

7.5.5 SUMMARY

- Process position and spread can be monitored using means and range charts for variables data and proportion defective and number of defects charts for attribute data.
- The interpretation of control charts involves the analysis of points

beyond the warning/action limits, trends in the data and any cyclical variation.

- Implementing control charts requires effective training, formalization of the controls and the establishment of charting parameters.

8 Problem-solving tools

8.1 The role of problem-solving

8.1.1 ROLE OF PROBLEM-SOLVING IN QUALITY DEVELOPMENT

To sustain the ongoing organizational development associated with 'improvement through teamwork' (Chapters 4 and 5) the improvement teams established need to be provided with appropriate tools. Nothing is more disheartening to an enthusiastic group of employees working on quality improvement than to have an inadequate approach with which to tackle the problem.

For all employees to have an understanding of the basic tools of problem-solving is seen by many organizations as a fundamental prerequisite on the journey to quality improvement. The importance of a structured approach to problem-solving is emphasized in an industrial context as many of the quality problems encountered are 'ragged'. Such problems inevitably have multiple causes, imprecise data and constrained solutions and require a problem-solving framework (methodology) and an appropriate set of tools (techniques).

To address these problems two sets of problem-solving tools have emerged. The first of these are the seven basic problem-solving tools, sometimes called the 'Seven QC Tools', which are normally used by quality improvement teams to effect improvement. The second set of techniques are the seven advanced tools usually referred to as the 'Seven Management Tools' which are used as advanced quality planning techniques. The role and application of each of these sets of tools in the quality development process is illustrated in Figure 8.1.

8.1.2 IMPROVEMENT VS INNOVATION

Problem-solving teams and problem-solving techniques are normally associated with an approach to quality development termed 'continuous improvement', often described using the Japanese word **kaizen**. The philosophy upon which kaizen is based aims at developing (often small-scale) improvements in all areas of the business on a continuous basis. It is the sum of all these individual improvements, for example, in:

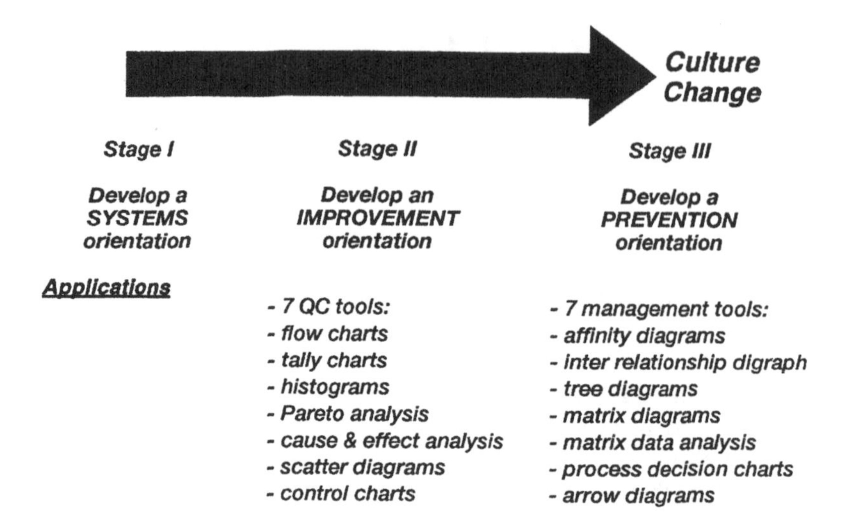

Figure 8.1 Application of problem-solving tools to quality development.

- reduced wastage of raw materials;
- reduced set-up times;
- better materials handling;
- reduced administrative inefficiencies;
- improved invoicing methods;

which over a period of time generates a significant cumulative business advantage. This kaizen or continuous improvement approach to quality development is characterized by:

- a large number of incremental improvements;
- based upon people involvement through teamwork;
- company-wide over a long period of time;
- originating from Japan.

This attention to improving the **detail** of every aspect of the work undertaken within the organization continuously over time is often cited as the underlying reason why Japanese industry achieved the spectacular manufacturing successes of the 1970s and 1980s.

Companies which have matured to a level of quality development characterized by kaizen have normally developed a passion for improvement throughout the organization. Such organizations achieve most of the improvements through teamwork using problem-solving techniques such as the 'Seven QC Tools'. This approach to enhancing business performance through improvement is often contrasted with an alternative ideology of innovation. Improving performance through innovation is primarily the domain of managers who look for new methods, materials or technologies to gain advantage over international competitors. The innovation approach to quality development is characterized by:

- infrequent, large-scale advance in performance;
- management driven;
- technology based;
- originating from the USA.

The tools and techniques used by organizations who develop their quality performance through innovation are typically those requiring fundamental rather than incremental change such as business process re-engineering (section 5.5). Such innovation is usually accompanied by a technological advancement such as the application of information technology (for example, networked computer systems, new materials and advances in plastic packaging materials) or new processes (for example, the float glass process).

There has been a considerable amount of debate amongst quality management professionals concerning the merits of improvement- rather than innovation-based quality deployment strategies. The comments often made include the following.

- Individual innovation is part of the culture of the 'West' whereas team-based improvement is part of the culture of the 'East'.
- Incremental improvement is too slow a process to close the competitive gap.
- Being first with the innovation is often an expensive and painful experience.

In reality the real quality development challenge is not how to choose between improvement- or innovation-based strategies, but how to combine the benefits of both approaches to enhance business performance. Long-term competitive advantage is achieved through the application of both improvement and innovation as shown in Figure 8.2.

Most companies therefore need to understand and combine both the techniques associated with incremental problem-solving for process improvement and the less frequently used methods of innovative process redesign. The techniques can be complementary and applicable to different aspects of the organization's operation.

8.1.3 IDENTIFYING IMPROVEMENT OPPORTUNITIES

The identification of quality problems would at first consideration appear to be a rather obvious exercise in the application of common sense. Surely organizations know exactly what their main quality problems are and where they manifest themselves? This is not universally true for the following reasons.

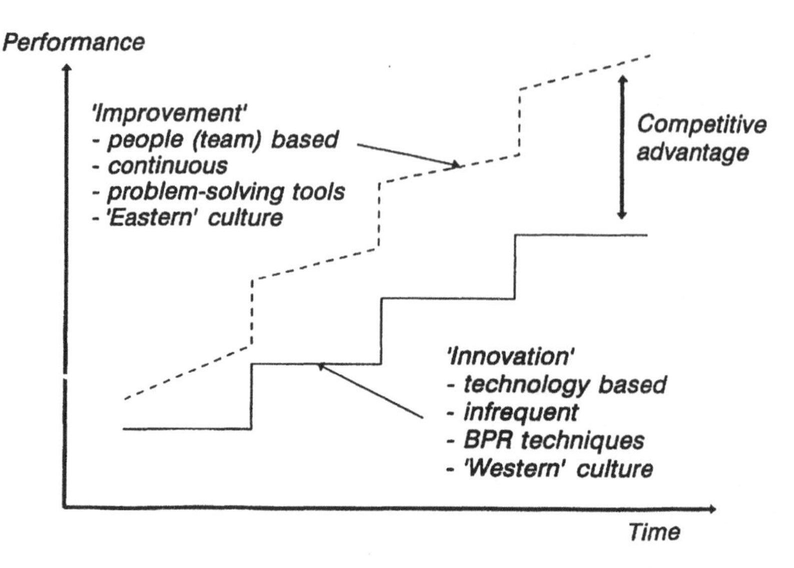

Figure 8.2 Improvement and innovation approaches to quality development.

- Very often it is the symptoms rather than the problems which are obvious to the organization.
- Very often quality problems experienced by the customer are not communicated to the organization.
- Very often problems are not prioritized in light of the quality objectives of the organization.
- Very often no mechanism exists to formally identify problems within the organization.

Before embarking upon a detailed problem-solving methodology, a company has to establish an approach to problem identification and these approaches are usually classified as either:

- **top-down**, deployed improvements; or
- **bottom-up**, self-identified improvements.

The type of improvement team established to tackle each of these problems is different (as described in section 4.3) both in terms of composition and approach adopted.

The **top-down** identification of quality improvement projects involves a process of deployment whereby the improvement objectives at a higher level are broken down to specific projects at a lower level. The stages in this process are as follows.

- **Stage 1.** Identify critical business objectives and from these identify the improvement objectives together with measures, goals and responsibilities.

- **Stage 2.** Take the improvement objectives identified in Stage 1 and decompose into the critical processes which need to be improved, and again measures and goals are identified.
- **Stage 3.** Continue the process described in Stage 2 until projects are 'bite-sized' and can be assigned to an improvement team to tackle.

This process of the top-down deployment of business objectives down to improvement teams is described above in section 4.3. Such improvement teams are described as CATs (corrective action teams) or QITs (quality improvement teams) and are normally made up from designated personnel within the organization from various functions. They are given a 'defined' time period within which to solve the improvement problem.

The **bottom-up** approach to problem identification utilizes the detailed knowledge of the individual team members to pinpoint opportunities for improvement. Such improvement teams, which are often referred to as **quality circles** or QC teams, are normally made up from volunteers within a particular function or department and have an ongoing brief to identify and improve quality problems.

The approach adopted by bottom-up improvement teams is as follows.

- **Stage 1.** Brainstorming activity by members of the team to identify quality problems within their function or department.
- **Stage 2.** Prioritize the problems identified in Stage 1 according to the importance perceived by the team and the probability of solving the problem.
- **Stage 3.** Tackle each of the quality improvement projects in priority order ensuring the team is responsible for implementing the improvements recommended.

Both of these alternative approaches to problem identification are appropriate to an overall programme of quality development and most organizations practising total quality will employ some combination of the two methods.

The major advantages and disadvantages of each approach are contrasted in Table 8.1. Although the mechanisms for identifying the problems differ, the tools and techniques employed are largely the same. Both types of quality problems require careful unravelling before solutions are identified.

8.1.4 SUMMARY

- Problem-solving methods and techniques are essential in assisting the contribution of improvement teams in the quality development process.
- Problem-solving tools contribute to an improvement strategy for quality development and for most organizations this should also be combined with appropriate innovative strategies.

Table 8.1 Advantages and disadvantages of the alternative project identification methods

Advantages	*Disadvantages*
Top-down	
Focuses clearly on the business objectives.	Relatively small numbers of people involved.
Requires little culture change to implement.	Does not create significant culture change at all levels in the organization.
Bottom-up	
Involves a wide range of people within the organization in the improvement process.	Problems identified can be trivial in terms of their contribution to the business.
Creates culture changes through employee empowerment.	Can create resourcing problems in terms of implementing the recommended solutions.

- Problems can be identified top-down through the deployment of quality objectives or bottom-up through prioritization by the individual improvement teams.

8.2 Problem-solving methodologies

8.2.1 QUALITY IMPROVEMENT METHODOLOGIES

As with most disciplines, the tools and techniques associated with quality management are most effective if used within some form of framework or methodology. Solving quality problems is not simply common sense. Problem-solving benefits from a structured approach in which the individual techniques are applied in an appropriate sequence and to an appropriate aspect of the problem.

The developments in quality management throughout the 1980s and 1990s have seen a greater emphasis upon the systematic collection, analysis and review of data particularly as organizations move from a 'systems' orientation towards an 'improvement' orientation.

A number of alternative and complementary improvement methodologies have emerged, the most widely publicized of which have been proposed by American quality 'gurus':

- the Deming cycle;
- the Juran trilogy;
- the Crosby process.

The cycle of continuous improvement proposed by Edwards Deming was originally developed by Walter Shewhart whose work at Bell Telephone Laboratories of Statistical Sciences particularly during the 1930s was fundamental to many of the subsequent quality improvement techniques. The Shewhart cycle of 'Plan–Do–Check–Act' was revised by Deming to include 'Study' rather than 'check' as shown in Figure 8.3.

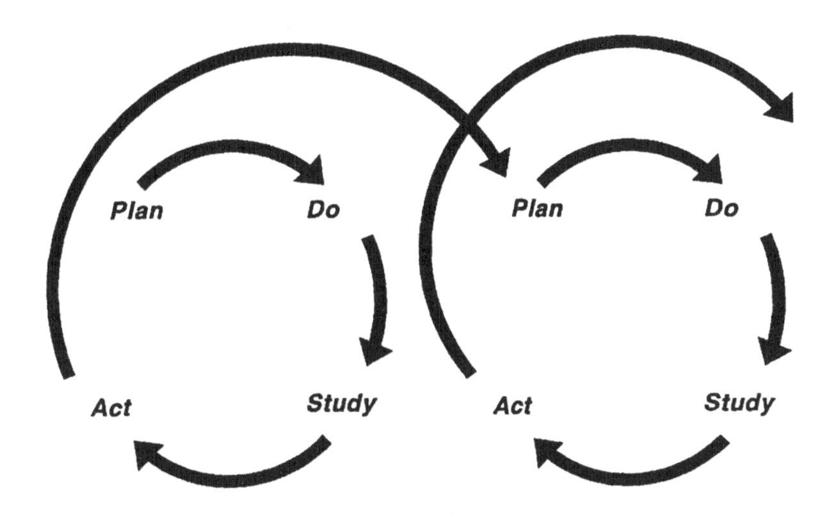

Figure 8.3 The Deming cycle PDSA.

This four-stage methodology is seen as a continuous process which utilizes the seven QC tools identified below in section 8.3.1. The process commences with the planning stage in which the current situation is analysed, data collected and plans made for improvement. This unravelling of the problem involves basic tools and techniques and reviews the effects of the actions taken in the previous cycle. The doing stage usually involves some form of trial or pilot solution, for example with a small part of a manufacturing or service process or small group of customers. This trial is critically evaluated during the study stage of the cycle and problems or other opportunities are examined. Finally the act stage ensures that the improvement is implemented in a standardized and continuous manner before embarking upon the next plan. The focus is therefore continuous and closed-loop improvement.

Juran, like Deming, emphasized the need for continuously working upon quality improvements and encourged the concept of 'breakthrough' whereby organizations would achieve improvements leading to new and unprecedented levels of performance as shown in Figure 8.4.

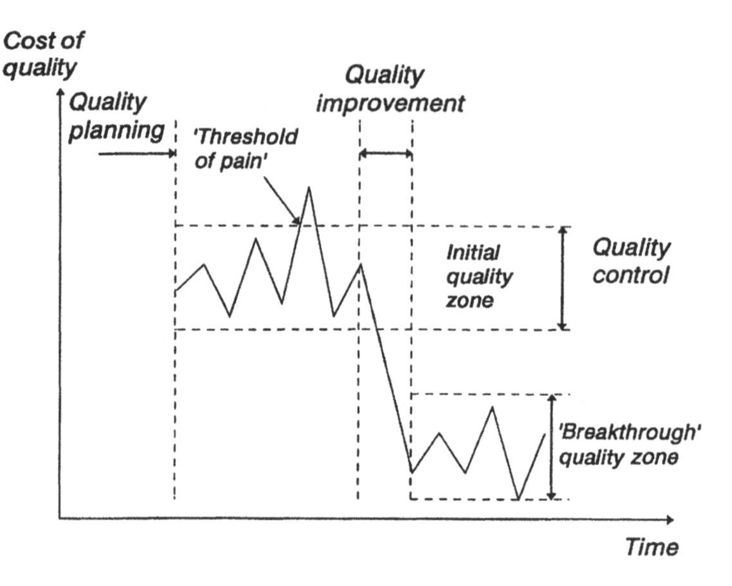

Figure 8.4 The Juran trilogy.

The Juran trilogy comprises the three stages of quality planning, quality control and quality improvement, and Juran suggested that most organizations placed too much emphasis on control and paid insufficient attention to the aspects of planning and improvement. The seven management tools support the quality planning process in assisting the identification and management of quality improvement opportunities. In terms of the quality improvement process, Juran proposed a six-stage methodology as follows.

1. **Proof of need** – to establish the economic benefit associated with quality improvement.
2. **Project identification** – to establish and distinguish the critical few from the trivial many improvement projects which may be undertaken.
3. **Organization for breakthrough** – the establishment of the project management and teamwork.
4. **The diagnostic journey** – the use of problem-solving tools to unravel the problem and identify the root causes of the problem.
5. **The remedial journey** – the selection of the optimal improvement proposal and effective implementation.
6. **Holding the gains** – ensuring that the new methods and procedures are established and the breakthrough in performance is maintained.

The Juran concept that quality improvement involves initially a diagnostic process to understand the problem followed by a remedial journey to identify solutions is fundamental to effective problem-solving. Juran also clearly identified the quality improvement process as a key management responsibility.

Crosby's approach to quality improvement is based upon what he describes as the four 'absolutes' of quality, namely:

- **Definition** – understanding that quality means conforming to the requirements.
- **System** – the approach to be adopted should focus upon prevention rather than inspection.
- **Performance standard** – the organization should strive for zero defects rather than adopt acceptable quality (defect) levels.
- **Measurement** – the true measurement of quality is the cost of non-conformances (Chapter 3).

In terms of a quality improvement methodology, Crosby describes this as a process rather than a programme to emphasize the continuous nature of improvement. The Crosby improvement process comprises 14 steps and is primarily a company-wide approach rather than an approach to solving an individual quality problem. The 14 steps are summarized in Table 8.2.

Table 8.2 Crosby's 14-step improvement process

Step	Process
1	Clear management commitment to the process.
2	Form quality improvement teams with representation across the organization.
3	Quality problem identification through effective measurement against which future improvement can be measured.
4	Evaluate the cost of quality to provide the company-wide measure of progress.
5	Quality awareness through the organization through increased communication.
6	Corrective action to resolve the quality problems identified utilizing problem-solving tools and techniques.
7	Establish an *ad hoc* committee to coordinate the 'zero defects' programme.
8	Train supervisors in the quality improvement process.
9	'Zero defects day' to communicate the concept to all employees.
10	Goal setting by the groups to establish the measurable improvement objectives.
11	Problem cause removal by encouraging the communication of problems and inhibitors to management.
12	Recognition.
13	Quality councils to coordinate and communicate progress and improvements to the process.
14	Do it over again after around 18 months to rejuvenate the process.

Each of the methodologies proposed by Deming, Juran and Crosby emphasizes the need for:

- a structured framework;
- the use of problem-solving tools;
- the measurement of improvement;
- the process to be continuous.

Many organizations have adopted these approaches proposed by quality 'gurus' only to find that they need to be tailored to meet company-specific needs. Increasingly as companies have matured and developed their improvement orientation they have adopted their own approach to quality problem-solving within the organization.

8.2.2 THE STAGES TO PROBLEM-SOLVING

The most effective problem-solving methodology to adopt will, not surprisingly, depend upon the type of quality problem to be tackled and upon the organizational mechanism to be used.

The philosophies of Deming, Juran and Crosby and others and the experiences of implementing quality improvement have established six basic fundamentals of problem-solving:

1. Problem-solving requires a structured approach.
2. Problem-solving requires clear measures of improvement.
3. Problem-solving requires the actual problem to be identified very clearly.
4. Problem-solving requires the problem to be effectively diagnosed and understood before the solutions are proposed and implemented.
5. Problem-solving is most effectively carried out by teams of people working together.
6. Problem-solving for quality improvement is a continuous process.

The benefits of adopting a step-by-step approach are twofold.

- At the outset to the project the improvement team has a basic game plan or route map through the problem which is important when tackling typical industrial quality problems.
- The stages of the problem-solving methodology can be used as a checklist against which the project's progress can be monitored and managed.

For these reasons, quality improvement problem-solving is normally undertaken utilizing a methodological approach with the basic stages shown in Figure 8.5.

This type of generic problem-solving framework can be used adaptively by many different types of organizations to solve a range of quality problems. Such an approach can assist in the facilitation of a wide range of improvement processes by providing the benefits of structure but allowing flexibility through the greater or lesser emphasis of any of the stages.

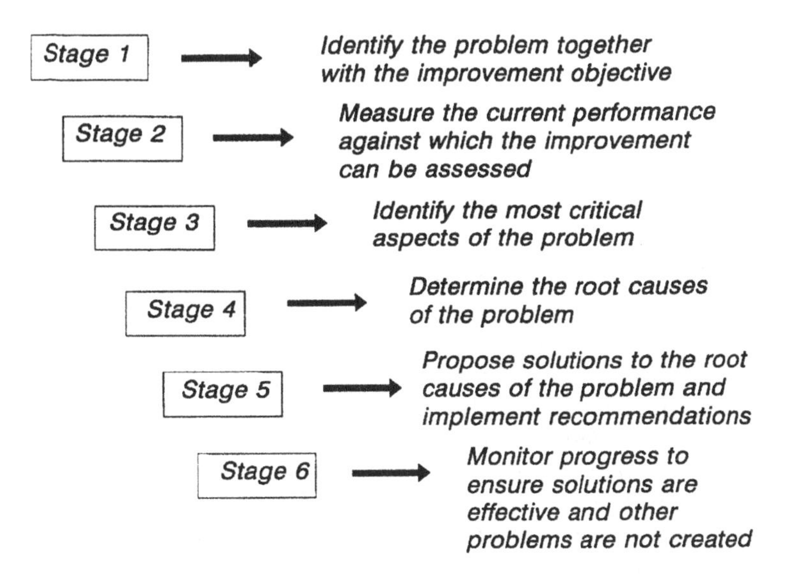

Figure 8.5 Basic stages in problem-solving.

Many organizations, teams and individuals will attempt to 'short-cut' the problem-solving procedure by, say, going directly from Stage 1 – problem identification – to Stage 5 – proposal of recommended solutions. This is a common mistake made as companies initially develop their improvement orientation. Quality problems need to be unravelled and clearly understood and therefore the discipline of undertaking both the diagnostic journey followed by the remedial journey is essential if robust, complete and effective solutions are to be found.

Different improvement tools and techniques are used at each of the stages of the methodology as discussed in section 8.3. The progress of a particular improvement team can be monitored against such a methodology and the final outcome of an improvement project can be conveniently reported in terms of the results and conclusions from each stage. This is particularly beneficial in assisting other people in the organization (for example, senior managers or employees affected by the quality improvement) to understand the basis upon which the quality improvement team have made their recommendations rather than simply being confronted with the 'solution'.

8.2.3 ALTERNATIVE METHODOLOGIES

As stated above, the most appropriate problem-solving methodology to adopt depends of course on the nature of the quality problem to be solved. The six-stage methodology proposed above in section 8.2.2 is robust and generic such that it can be used in a range of different applications in both manufacturing and service organizations. A distinction does emerge, however, when applying problem-solving to the type of improvements described in section 8.1.3, namely:

- top-down improvement problems deployed down from business/quality objectives;
- bottom-up improvement problems identified by the improvement teams themselves.

The key differences in terms of the problem-solving methodology are in the application of the first three stages.

In the case of the deployed improvement project the important element of Stage 1 is to ensure that the problem is clearly understood by the team and that the improvement goal is 'owned' by the team members. The project brief may be amended following the analysis of Stage 3 to more tightly bound the problem or alternatively to update the current performance level.

For the self-identified quality problem the first three stages of the methodology are often iterative with the improvement team identifying a range of problems in Stage 1 and then prioritizing the problems through the measurement and pinpointing of Stages 2 and 3.

It is often useful therefore in the implementation of quality improvement for a problem-solving team to undertake an interim presentation to the quality steering group (or council) after Stage 3 to ensure the problem is either understood (top-down) or appropriate (bottom-up).

8.2.4 SUMMARY

- The process of quality improvement is a continuous cycle that requires a structured approach or methodology.
- The key stages in problem-solving are the identification of the problem, the measurement of the current performance, the pinpointing of critical areas, the determination of root causes, the recommendation of solutions and the tracking of progress.
- Problem-solving needs to be adaptable to accommodate both top-down and bottom-up improvements.

8.3 Basic problem-solving tools

8.3.1 THE SEVEN BASIC PROBLEM-SOLVING TOOLS

To support the step-by-step problem–solving methodologies described in section 8.2, quality improvement teams also require a basic set of analytical techniques which will assist in the measurement, understanding, pinpointing and solving of the problem. A basic set of seven problem–solving tools, was proposed by the Japanese quality 'guru' Ishikawa as a fundamental 'toolkit' which should be understood by all employees involved in quality improvement. These basic tools are:

- flow charts;
- tally charts/checksheets;
- histograms/stratification;
- Pareto analysis;
- cause and effect analysis;
- scatter diagrams;
- control charts/run charts.

All of these techniques are not necessarily required to tackle every problem and in certain situations more sophisticated tools are needed to analyse complex data variation. The seven basic tools do, however, represent a useful, general set of techniques to support company-side quality improvement programmes.

The first of these tools, **flow charting**, is a systematic technique for describing the process to which the problem belongs. The main purpose for producing a flow chart is to establish a common understanding of all the stages of the process under review. Very often the process of generating the flow chart will require a number of different insights from different team members and therefore represents an excellent vehicle for communication of the problem. On many occasions the simple act of producing an accurate flow chart of the problem will identify unnecessary or critical stages where the subsequent improvements may be focused. Figure 8.6 shows a typical flow chart generated by a quality improvement team which identified a

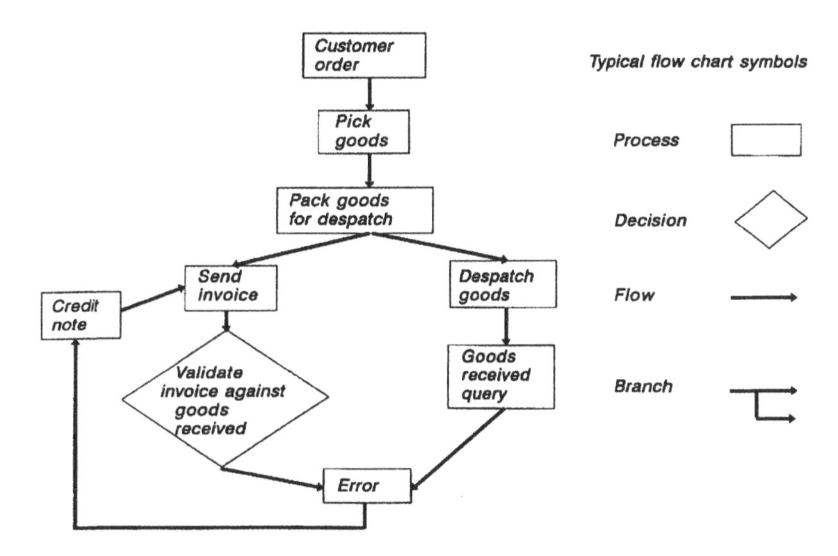

Figure 8.6 Flow chart.

major problem with invoicing errors in a service company caused by the issue of invoices against despatched goods rather than confirmed deliveries.

Tally charts or Checksheets are used to assist in the collection and classification of data. In many manufacturing process applications it is essential to plan the data collection exercise carefully to facilitate accurate (and rapid) information gathering. Very often some form of brainstorming exercise is required from the quality improvement team prior to the data collection exercise to identify the main categories into which the data will be classified. A typical tally chart for the collection of a monthly set of customer complaints is shown in Figure 8.7.

Histograms are used to display the data collected and often this form of ordered distribution illustrates fundamental properties of the problem. Patterns observed in the data when presented in the form of a histogram are often indicative as to the nature of the problem, for example certain classes of data dominate or the data is skewed towards certain data values. Graphical representations of the data using histograms will often contradict the 'myths' which arise when individuals believe they understand the main causes of a problem which is not substantiated by the data (facts). The information presented in a histogram can often be more informatively presented if further **stratification** is undertaken. Stratification further separates the data according to dimensions such as:

- process type or location;
- material type or supplier;

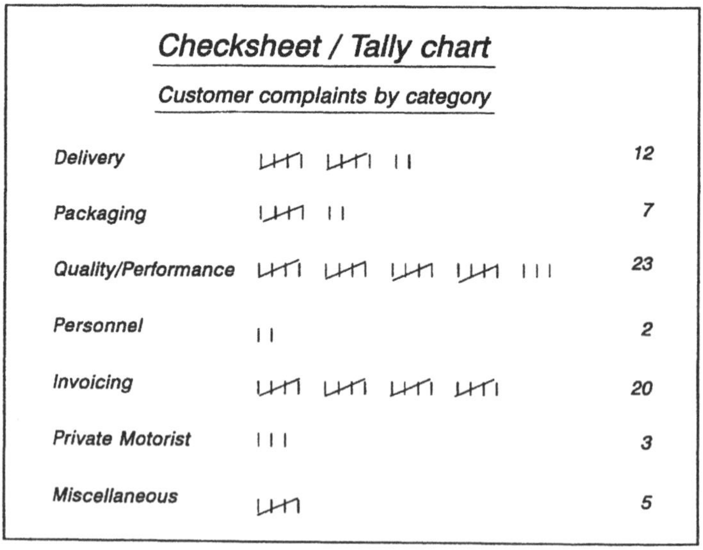

Checksheet / Tally chart

Customer complaints by category

Category	Count
Delivery	*12*
Packaging	*7*
Quality/Performance	*23*
Personnel	*2*
Invoicing	*20*
Private Motorist	*3*
Miscellaneous	*5*

Figure 8.7 Tally chart/checksheet.

- time of day or year;
- operator experience or shift;
- environmental conditions.

A histogram for the data collected above is shown in Figure 8.8 together with further stratification of the data.

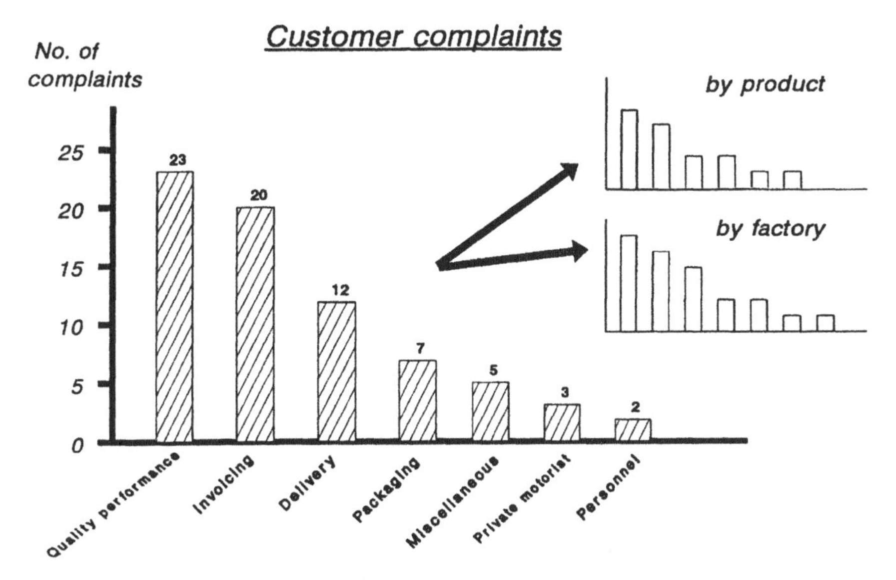

Figure 8.8 Histogram with stratification.

Pareto analysis is an approach to pinpointing problems through the identification and separation of the vital few from the trivial many. The technique is named after a nineteenth-century Italian economist and social scientist, Vilfredo Pareto, who examined the distribution of wealth in Italian society and found that a relatively small number of people (around 10%) owned a proportionately large amount of the wealth (around 90%). This rule of diminishing return is exhibited by many complex systems and therefore represents a useful insight for industrial problem-solving. In business this same concept is often referred to as the 80:20 rule whereby:

- 80% of the revenue usually come from 20% of the products in the range;
- 80% of the problems are normally due to 20% of the causes.

The objective in many problem-solving situations is to 'sort the wheat from the chaff', in other words to identify the critical (fewer) elements which are contributing to the majority of the problem. The process involves ordering the histogram according to frequency and then plotting the cumulative frequency as shown in Figure 8.9.

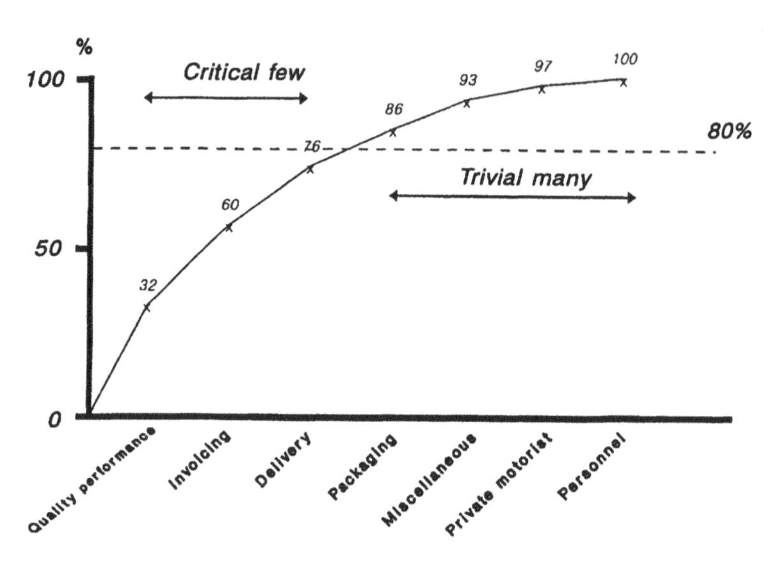

Figure 8.9 Pareto analysis.

From such a chart, the critical elements of the problem can be identified by drawing a line across at around the 80% cumulative level. In terms of problem-solving, the quality improvement team can use this approach to focus their attention on the relatively small number of factors/causes which contribute most to the problem. If the Pareto chart does not exhibit a cumulative plot of diminishing return then this often indicates that the factors have not been correctly stratified or grouped.

Cause and effect analysis is used as a way of structuring the creative process of determining the root cause of a problem. This technique utilizes a cause and effect diagram (sometimes also called a fishbone diagram due to its appearance or an Ishikawa diagram after Dr Ishikawa who originally applied this technique in Japan) in which the problem or effect is drawn at the end of the main horizontal stem. Each of the causes of the problem are then drawn as branches to the main stem onto which are drawn the contributory causes. The causes are often classified according to the seven Ms, namely:

- Machine (processes);
- Materials (inputs);
- Methods (procedures);
- Measurement (errors);
- Manpower (people);
- Money (constraints);
- Milieu (environment or culture);

although this is not the only classification of causes to adopt as other groupings (such as by product type, by customer or by location) may be more helpful to the creative brainstorming process.

These diagrams are normally generated as part of a team-based brainstorming exercise in which all the members of the team are encouraged to contribute ideas on the possible causes of the problem. By drawing the cause and effect diagram a 'rich picture' of the problem is created to which subsequent solutions can be more readily attached. A typical cause and effect diagram is shown in Figure 8.10 and this technique is often used iteratively with Pareto analysis to identify and pinpoint causes to quality problems.

Scatter diagrams are a technique used to establish (or dispel) a causal link between two factors and often having established some correlation the solution to the problem becomes very much easier to identify. Such diagrams can illustrate:

- positive correlation;
- negative correlation;
- curvilinear correlation;
- weak correlation;
- no correlation.

The scatter diagrams are drawn as X–Y plots in which the data for a number of different instances are shown on the diagram. In industrial situations the data is often not entirely accurate and the correlations often weak; however, even in such cases, the scatter diagram can indicate a way forward to solving the problem. For example, establishing a positive relationship

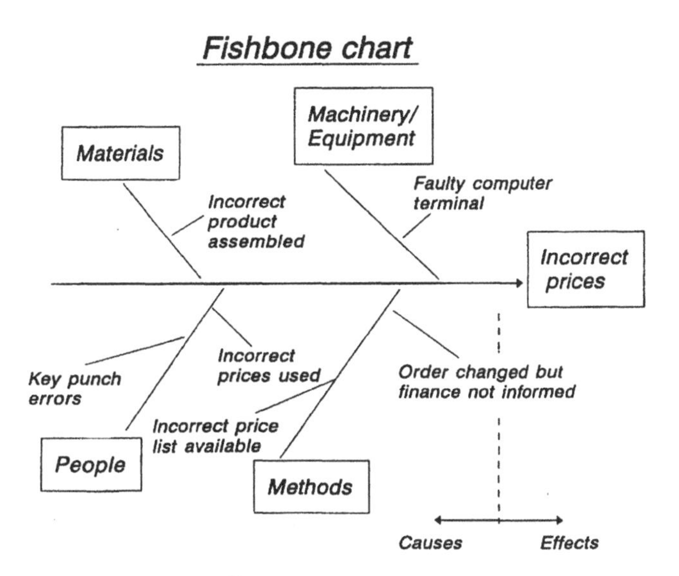

Figure 8.10 Cause and effect diagram.

between the accuracy of demand forecasting for a product and the number of occasions, say, the product is promoted (price reductions, advertising campaigns) helps to focus the improvement efforts towards solving the problem of forecast error. Typical scatter diagrams are shown in Figure 8.11.

Control charts are basically a form of run chart onto which control limits are drawn to indicate when the process is behaving 'normally'. If points on the chart fall outside these control limits then the process is being influenced by special (non-random) factors. The development and construction control charts is described more fully in Chapter 7.

The control limits are normally set at plus and minus three standard deviations of the data and therefore indicate a probability of around 99.7% that points outside these limits have not occurred naturally. By plotting the data in this way, the quality improvement team can observe whether the process under consideration is being affected by significant external factors or whether any trends are emerging in the data. A very useful associated technique in the context of problem-solving is the calculation of process capability C_p (described in detail in section 7.4). This represents the ratio of the tolerance allowed for the process to the effective range of the process (six standard deviations). If this ratio is greater than unity the problem-solving team can deduce that the process is capable of delivering the required quality and perhaps focus their attention on those processes or aspect of the process where the process capability is less than one. Identifying where

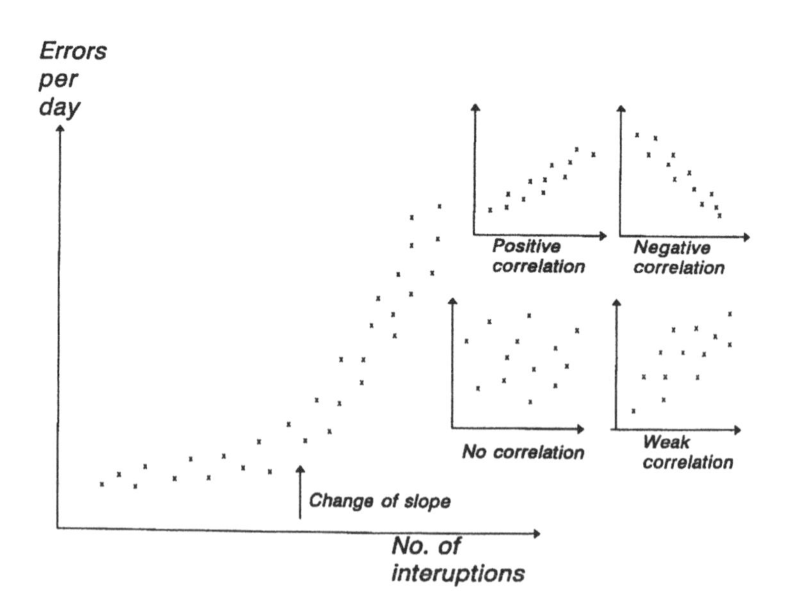

Figure 8.11 Scatter diagrams.

process capabilities are poor is a useful tool for pinpointing where the potential sources of quality problems may exist.

A typical problem-solving application of a control chart is shown in Figure 8.12 together with the effects of ranging process capability.

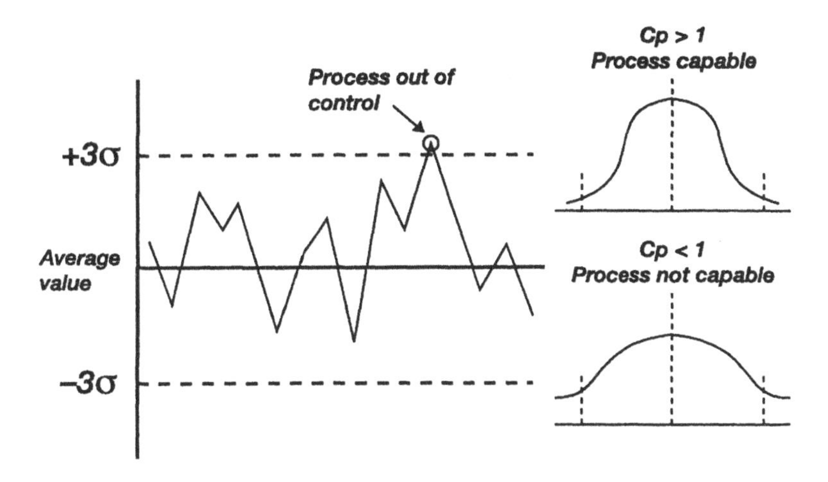

Figure 8.12 Control charts and process capability.

8.3.2 THE APPLICATION OF PROBLEM-SOLVING TOOLS

Each of the seven basic problem-solving tools can be used within the quality improvement methodology described in section 8.2.2. Some of the techniques can be used at more than one stage in the methodology and certain groups of techniques (particularly those used for pinpointing and then evaluation) can be used iteratively within a given stage.

The application of the problem-solving tools to the various diagnostic stages of the methodology is shown in Figure 8.13.

Stage 5 of the methodology, the proposal and implementation of solutions, is problem-specific and requires the experience and creative insights of the team members. If, however, the problem has been adequately analaysed and understood this process of identifying solutions becomes much simpler and more effective.

The application of the tools within the methodological framework is also enhanced through the guidance of a team facilitator as discussed in section 4.4. The problem-solving methodology should be used by the facilitator to guide the team through the problem and in addition the facilitator should be experienced in the application of the individual tools and techniques and encourage team members to use them appropriately.

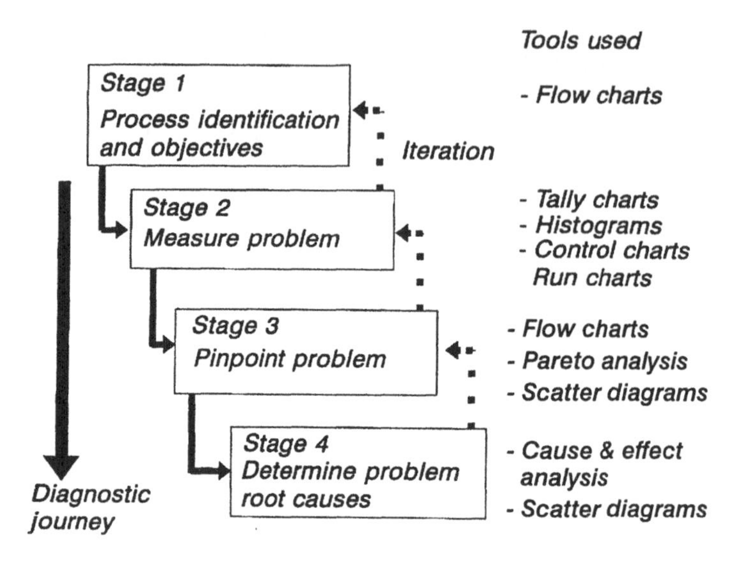

Figure 8.13 Application of problem-solving tools.

8.3.3 PROGRESS TRACKING

In the identification and implementation of solutions to quality problems, it is very easy to convince oneself that the problem has been eliminated through a creative and thorough piece of problem-solving. In reality often the root causes can be more complex and difficult to eliminate and therefore for problem-solving to be rigorous, emphasis should be given to monitoring the effectiveness of the solution. This monitoring is normally termed **progress tracking**. The two important techniques used for progress tracking are:

- run charts (often multi-level);
- cusum charts.

Monitoring the effectiveness of the solution to a quality problem using a **run chart** is the most basic form of validating the improvement. It is often advisable that this monitoring is continued for a reasonable length of time to ensure the improvements are held and not simply transient due to a 'Hawthorne effect' (the effect on an activity due simply to being observed) from the attention to the problem given by the quality improvement team. In addition multi-level tracking should be employed where appropriate to ensure that progress in one aspect of a problem (for example, delays in the sending out of invoices) is not made at the expense of a different aspect (for example, invoice accuracy), as shown in Figure 8.14.

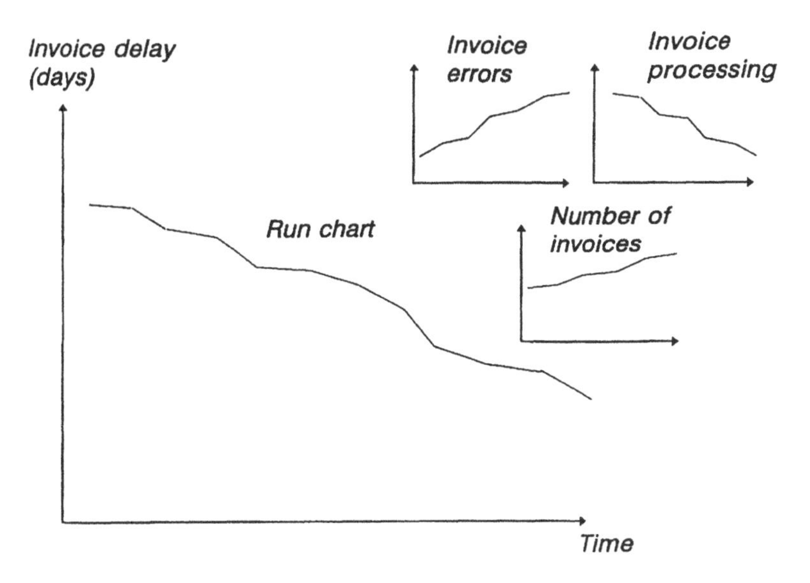

Figure 8.14 Multi-level progress tracking.

Where the quality improvement produces relatively small changes in the average value then **cusum charts** are an extremely useful technique to identify such variations. The cusum or cumulative sum is calculated by summing the differences between the actual value of the variable under evaluation and the target value. When this cumulative sum is plotted on a cusum chart then an increasing average results in a very obvious positive gradient and similarly a decrease in average will produce a negative gradient.

The usefulness of cusum charts in the context of problem-solving is that they provide information very quickly to the problem-solving team that a particular improvement is having an effect. The method for calculating the cumulative sum and the plotting of the chart is shown in Figure 8.15.

8.3.4 SUMMARY

- The seven basic problem-solving tools necessary for quality improvement are flow charts, tally charts, histograms, Pareto analysis, scatter diagrams and control charts.
- Each of the seven basic tools can be used within the problem-solving methodology to assist in the diagnostic journey to unravel the problem.
- Proposed solutions to quality problems need to be monitored to ensure the improvements materialize and are sustained.

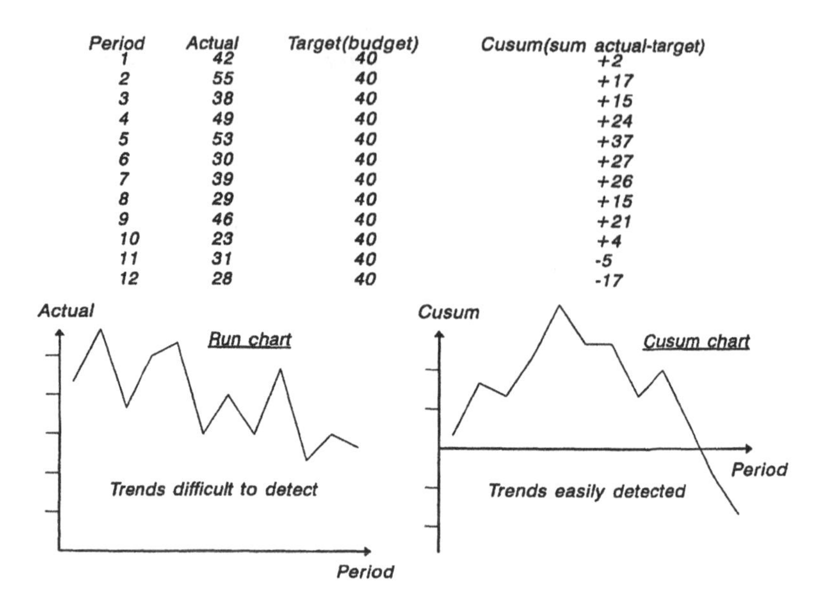

Period	Actual	Target(budget)	Cusum(sum actual-target)
1	42	40	+2
2	55	40	+17
3	38	40	+15
4	49	40	+24
5	53	40	+37
6	30	40	+27
7	39	40	+26
8	29	40	+15
9	46	40	+21
10	23	40	+4
11	31	40	-5
12	28	40	-17

Figure 8.15 Cusum charts.

8.4 Advanced problem-solving tools

8.4.1 THE SEVEN MANAGEMENT TOOLS

As organizations mature in the way quality is managed (Chapter 1) so they move away from solving problems (improvement) and focus more towards eliminating problems (prevention). A critical element of developing a prevention-orientated approach is a greater emphasis upon quality planning and the focusing of management attention onto the design and implementation of effective products, services or processes rather than the improvement of defective ones. The techniques for advanced quality planning are discussed in Chapter 10 and in terms of advanced problem-solving tools the appropriate techniques are usually referred to as the seven 'new' or 'management' tools. These advanced tools are:

- affinity diagrams;
- interrelationship digraphs;
- tree diagrams;
- matrix diagrams;
- matrix data analysis;
- process decision programme charts;
- arrow diagrams.

As with the basic problem-solving tools, not all of these techniques are required in every problem-solving situation and they are best applied by managers working together in teams.

The first of these advanced tools **affinity diagrams**, is an approach to simplifying complex problems through the systematic grouping of the many ideas generated during a brainstorming session. The basic process of creating an affinity diagram involves:

- generating a broad statement of the problem to be considered by the team (normally comprising six to eight members);
- each team member generating as many ideas as is possible relating to the problem, these being recorded on individual cards;
- the team then grouping the ideas together into logical sets each of which is then given a generic description which summarizes the essence of the idea.

This technique helps to bring some form of hierarchical structure to the complexity of a problem and therefore assists in the 'mapping' of the problem. A typical affinity diagram is shown in Figure 8.16.

An **interrelationship digraph** describes the logical links among the factors grouped together during a brainstorming session such as produced by the affinity diagram. By focusing upon a particular idea (generated, for example, by an affinity diagram) the interrelationship digraph brings logical structure and relationships to the creative activity. The nomenclature used by the interrelationship digraph describes processes, causes and results and

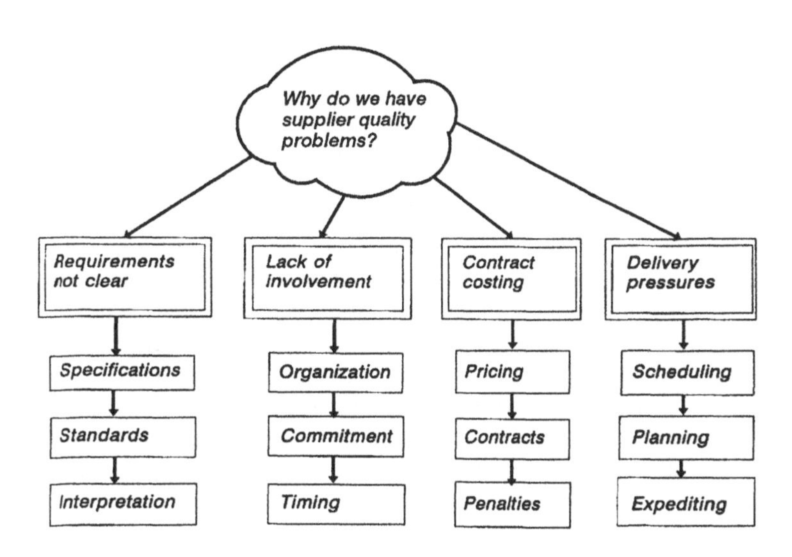

Figure 8.16 Affinity diagram.

is therefore a more powerful version of the Ishikawa diagramming technique (section 8.3.1). Again the digraph is generated by a team and is helpful in the mapping out of the logical relationships which exist within a problem as shown in Figure 8.17.

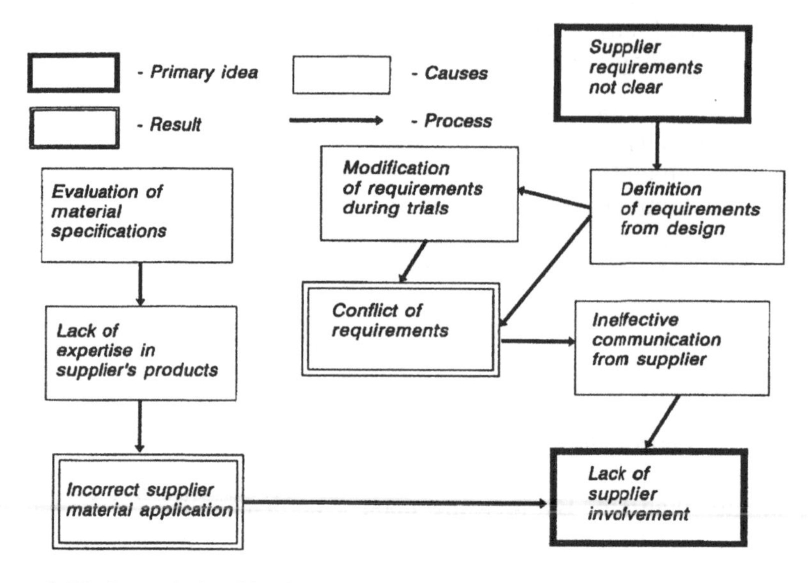

Figure 8.17 Interrelationship digraph.

A **tree diagram** is used where the sequential structure of a problem needs to be understood or where the stages to implementing a solution need to be established. This type of diagram is useful where customer requirements need to be understood more fully and can be broken down to successive levels of detail in a sequential order. Tree diagrams assist the team in understanding the sequence of events which either cause the problem or are required to effect a solution. A typical tree diagram is shown in Figure 8.18.

The **matrix diagram** is a two-dimensional array similar to a spreadsheet which graphically represents the relationship between factors. The most commonly used example of a matrix diagram is the 'House of quality' diagram developed as part of quality function deployment described in section 10.2. This technique is particularly useful where factors used in different parts of an organization need to be brought together to facilitate improvement. For example, a matrix diagram can be used to relate the requirements of customers as described by the marketing department (appearance, visual impact, type, etc.) to the specifications of the product as described by the design or technical departments. An 'L' shaped matrix diagram could be used by the team to relate these two sets of factors or a 'T' matrix used to bring in the relationship with a third factor, for example the performance of a

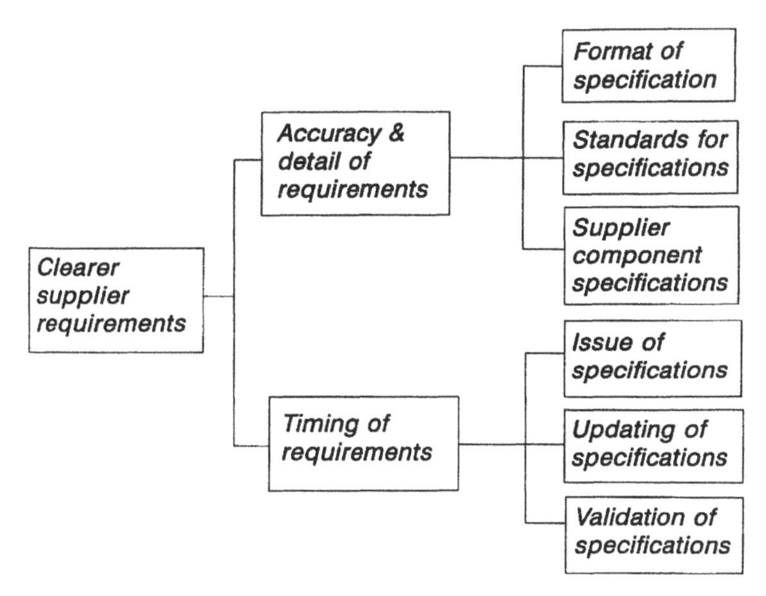

Figure 8.18 Tree diagram.

competitor's product. Figure 8.19 shows the application of matrix diagrams. The only one of the seven management tools which is used for analysing data is **matrix data analysis**. This technique is used to evaluate the numerical weighting of the relationships identified in a matrix diagram. The technique employs factor analysis to prioritize the respective correlations between the relationships and because of its quantitative nature it is perhaps the least used of the seven tools. Matrix analysis is useful, however, in quantifying product or service factors in terms of their preference by customers in the marketplace. It is therefore used primarily by product marketing to 'position' the features of the product in relation to certain customer factors as illustrated in Figure 8.20.

The technique of **process decision programme charting** is a method for mapping out all the stages and contingencies in going from the problem statement to the problem solution. The key to producing a process decision programme chart (PDPC) is to anticipate all the possible failures or problems which may occur and to therefore anticipate and plan for such problems. The technique is the problem-solving graphical equivalent to failure mode effect analysis described in section 7.2 and typically the PDPC is drawn to represent a branch from the tree diagram. The chart produced is very much a quality planning tool used to 'think through' the implications of a particular problem solution. At each stage in the preparation of the chart, the team is required to consider: 'What problems could arise at this point?' The resultant chart can then be used to plan contingencies for each of the major problems anticipated as shown in Figure 8.21.

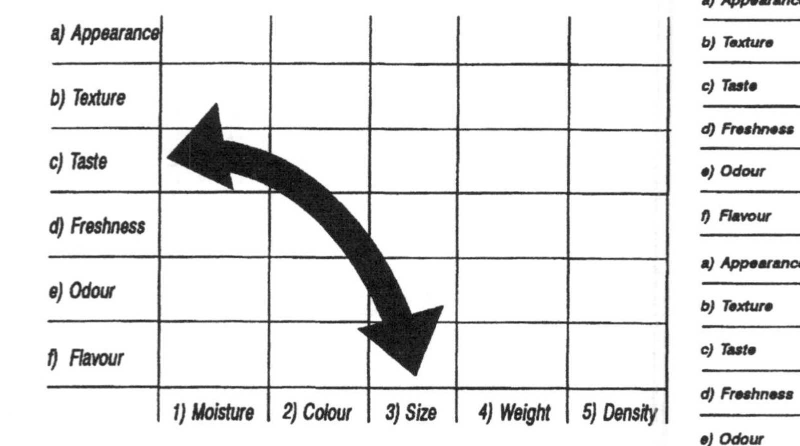
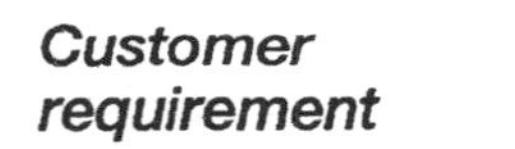

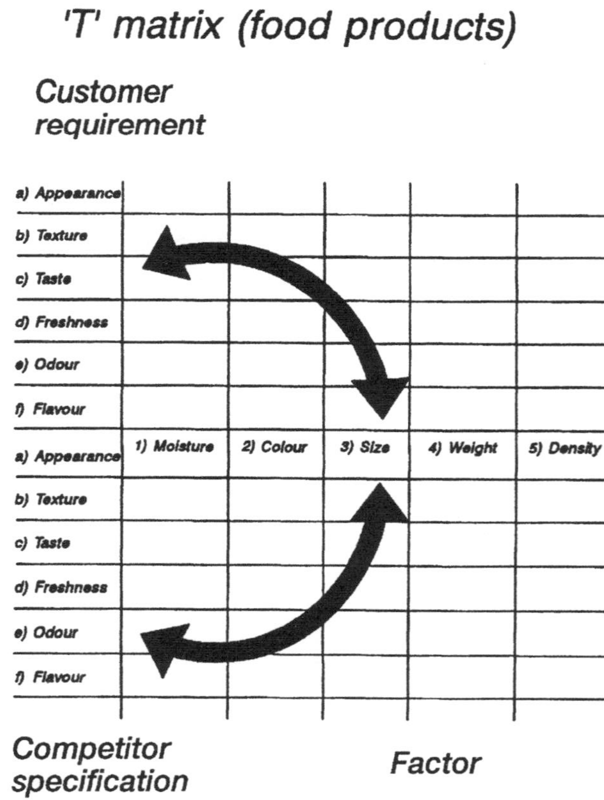

Figure 8.19 Matrix diagrams.

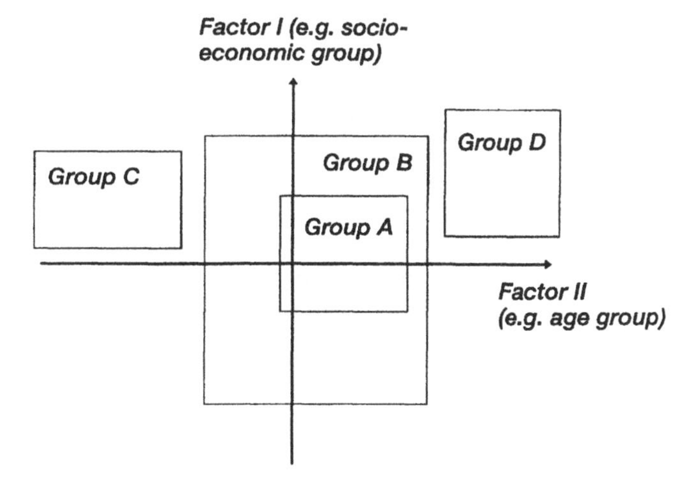

Figure 8.20 Matrix analysis.

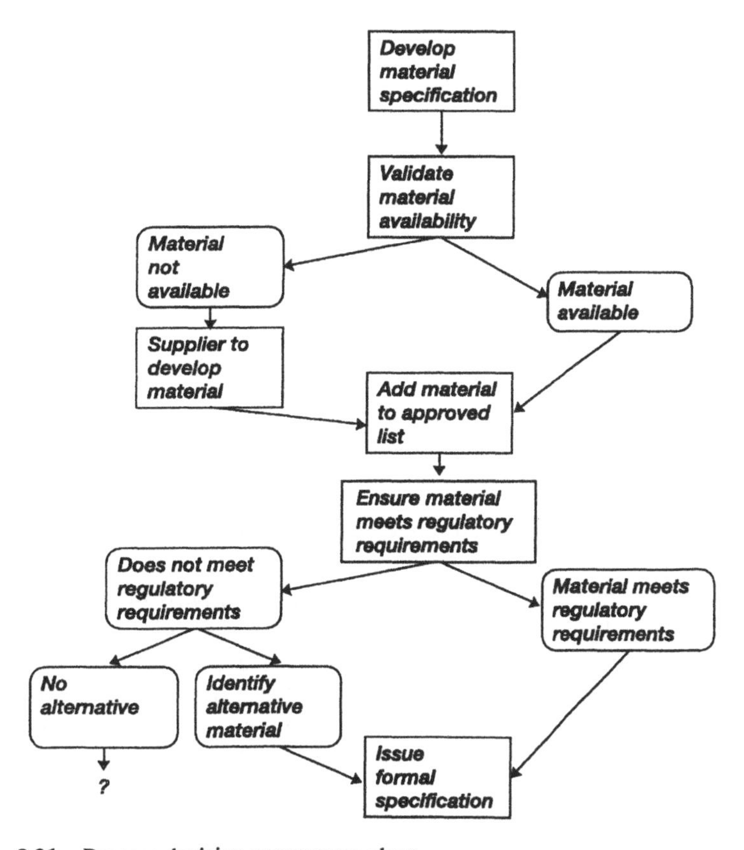

Figure 8.21 Process decision programme chart.

The last of the seven management tools is the **arrow diagram** which is used to systematically plan and schedule the quality developments. The tool is equivalent to the widely used project planning technique, programme evaluation and review technique (PERT) and is basically a diagramming method for illustrating the sequence, precedence and duration of events. The arrow diagram is therefore a more informative form of Gantt chart and is primarily used as a technique for both understanding a problem and planning the implementation of the solution. Arrow diagrams assist in the identification of critical project paths and also the overall likely project implementation timescale. A typical arrow diagram is shown in Figure 8.22.

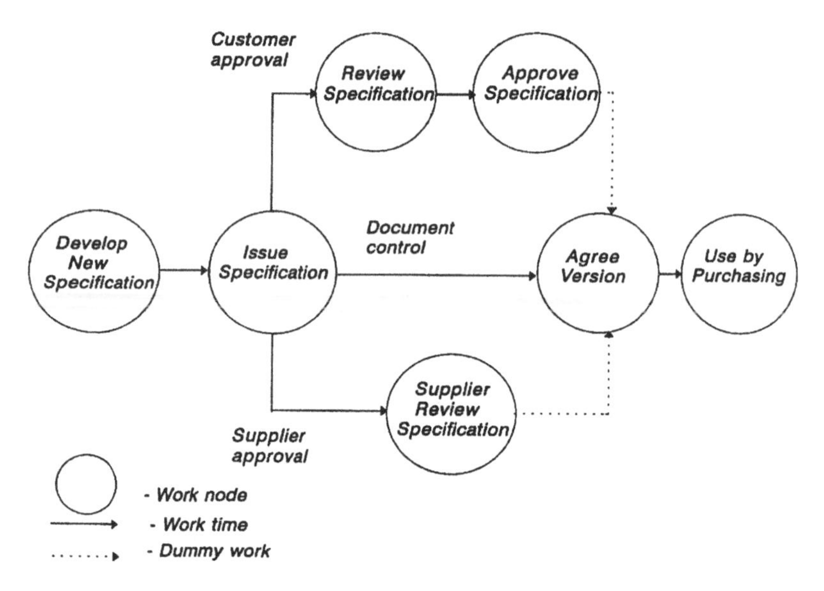

Figure 8.22 Arrow diagram.

8.4.2 THE APPLICATION OF ADVANCED PROBLEM-SOLVING TOOLS

As with the seven basic tools, the advanced problem-solving techniques need to be applied selectively within an overall framework of quality development. Primarily the seven management tools are used in the design of products, services or processes and ensuring these are free from problems. They are primarily concerned with improvement prior to production or execution and are therefore associated with quality planning and a prevention-orientated approach.

The seven management tools can be used to supplement the basic problem-solving activities particularly for diagnosing complex (often organization-wide) problems and for assisting in the planning of the implementation of solutions. In terms of the basic problem-solving methodology described in section 8.2.2 the advanced tools can be applied as illustrated in Figure 8.23.

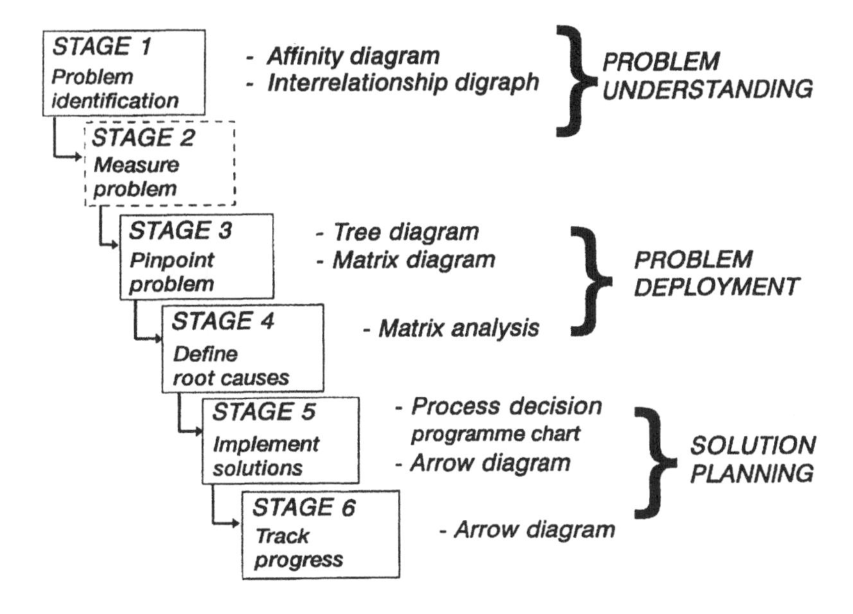

Figure 8.23 Application of the seven management tools.

In addition these tools are used as part of team-based advanced quality planning techniques such as quality function deployment and Taguchi analysis as described in Chapter 10. Many mature quality developed companies are increasingly using the seven management tools as techniques within their own design project planning methodologies for both the systematic capture of customer requirements and for structuring project management activity.

8.4.3 SUMMARY

- The seven 'new' management tools for advanced problem-solving are affinity diagrams, interrelationship digraphs, tree diagrams, matrix diagrams, matrix analysis, process decision programme charts and arrow diagrams.
- These advanced tools are associated with improved quality planning and the implementation of solutions to more complex quality problems.

9 Reliability management

9.1 Product and systems reliability for quality improvement

9.1.1 THE ROLE OF RELIABILITY IN QUALITY DEVELOPMENT

The reliability of a product or service is synonymous with the concept of 'Quality' in the mind of the customer. Reliability is perceived to be the quality performance of the product or service over time, and in many sectors of industry and commerce reliability is considered the most important attribute. For example:

- in safety-critical applications;
- in military applications;
- in transport applications.

The concept of maintaining product or service quality over time is very much part of the development of an organization in the way in which quality is managed.

In the early stages of development, where the focus is upon the implementation of effective systems, the critical activity is the defining of performance requirements and establishing procedures and methods to meet these requirements. As organizations mature, improvements are made to both the product and processes to enhance reliability and reduce the causes of failure. Finally as a prevention orientation dominates the reliability is incorporated into the basic design and the product focus is for failure proofing and the process emphasis is upon preventative maintenance.

The application of the tools and techniques of reliability management in the quality development process is shown in Figure 9.1.

Within manufacturing, the management of reliability focuses upon both:

- product reliability in terms of the performance in the marketplace and the perceptions of the customer;
- process reliability in terms of the availability of the manufacturing process and the consistency of production.

The business pressures to improve each of these elements are:

- product liability and the requirements of manufacturers to produce products which are safe in service and meet the growing national and

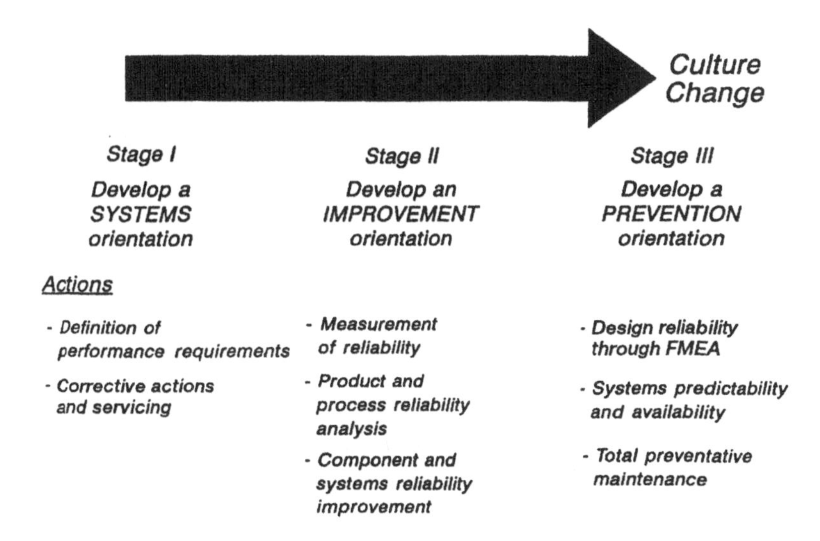

Figure 9.1 Application of reliability management to the quality development process.

international product liability legislation;
- manufacturing performance and process repeatability largely being determined by production reliability and availability, particularly in the process industries.

Certainly the techniques and concepts of reliability management are now an integral part of business development through quality.

9.1.2 BASIC ELEMENTS OF RELIABILITY

The key concepts in the management of reliability can be understood from the following definitions:

- **Reliability** – this is the probability that a product (or system) will perform its intended function satisfactorily for a stated period of time under specific operating conditions.
- **Probability** – the analysis of reliability involves the determination and manipulation of probabilistic values.
- **Functionality** – the operation of the product or system needs to be defined in terms of the requirements of the customer or the design specification.
- **Time** – the stated reliability is quoted at a specific point in time and represents the cumulative probability to that point in time.
- **Operating conditions** – the environmental conditions under which the product or system operates is clearly a critical factor in the reliability and needs to be defined and adhered to.

Two of these elements are concerned with measurement, namely probability and time, whilst the other two, functionality and operating conditions, are concerned with specification. Most of the analytical techniques associated with reliability management are therefore concerned with the relationship between probability and time. Reliability from the definition above is the probability that the component or system will still be functioning at the end of the stated time period and is therefore a survival function. Commonly used is the corresponding failure function and very often the performance measures associated with component or system reliability are descriptions of failure probability.

9.1.3 RELIABILITY MANAGEMENT AND COSTS

From a quality management perspective, the main organizational areas for the application of reliability techniques are:

- **design management** promoting failure mode analysis and the development of robust product designs;
- **maintenance management** ensuring the availability of processes and the ability to conform to planned arrangements and to minimize disruptive costs;
- **logistics management** to ensure the availability of products on time and to predict the requirements for spares and service provision;
- **marketing management** to evaluate the performance of products in the marketplace and in particular to assess the performance of new products or services.

The generalized techniques described below in sections 9.2 and 9.3 can be used in each of these areas of the organization and are increasingly applied to the planning activities within companies.

Evaluating the costs associated with product or process reliability is also an important contribution to quality development. The maturing company evaluates the mechanisms for the conversion of failure costs (product failures, equipment breakdown) to prevention costs (reliability and failure analysis). As described in Chapter 3, the process cost model can be used to evaluate the relationship between the cost of conformance (reliability) and the cost of non-conformance (failure).

In terms of the typical cost analysis for capital equipment, the lifecycle costs (LCC) are often calculated where:

$$\text{Lifecycle costs} = I + K\,(O + M + S) - R$$

where:
- I = initial costs of equipment;
- K = lifecycle constant;
- O = operating costs per annum;
- M = maintenance costs per annum;

S = standing costs per annum;
R = residual value of equipment.

This more 'total view' of the cost of equipment shows the dependency upon the reliability of the equipment through the cost of maintenance. By taking a simplistic view based upon the initial price of the equipment (I), the lifecycle profit (Lifecycle costs – Lifecycle revenues) is optimized by minimizing the total LCC and not simply by taking the option which affords the lowest initial or operating costs. The lifecycle costs associated with both process and product reliability are shown in Figure 9.2.

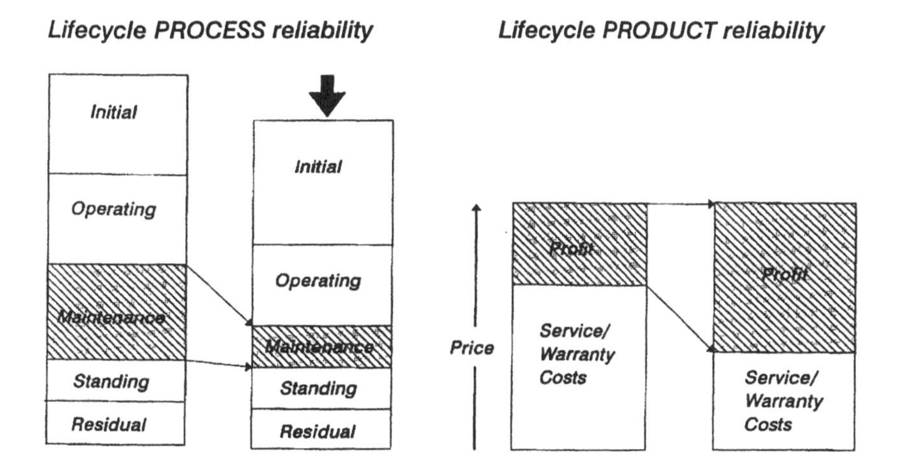

Figure 9.2 Lifecycle costs for process and product reliability.

9.1.4 SUMMARY

- Reliability represents the performance over time of a product or service and is an important parameter in all stages of quality development.
- The techniques of reliability management are aimed at improving both product and process performance with time.
- The business importance of reliability management is seen both in the reduced cost of failure through improved product design or process maintenance and also lower lifecycle and liability costs.

9.2 The modelling of product reliability

9.2.1 FAILURE CHARACTERISTICS

As described above in section 9.1.2 the definition of reliability is based upon the relationship between the failure probability and time. All of the characteristics used to describe and measure reliability are based upon this probability – time relationship.

The basic function describing the variation of the failure probability with time f(t) is termed the **probability density function** (pdf). The cumulative probability over a time period t is the integral of the pdf and is termed the **unreliability function** F(t) where:

$$F(t) = \int_0^{\infty} f(t).dt$$

Given that failure and survival (reliability) are complementary and mutually exclusive, then the **reliability function** R(t) is therefore:

$$R(t) = 1 - F(t)$$

The relationship between the probability density function, the unreliability function and the reliability function is illustrated in Figure 9.3.

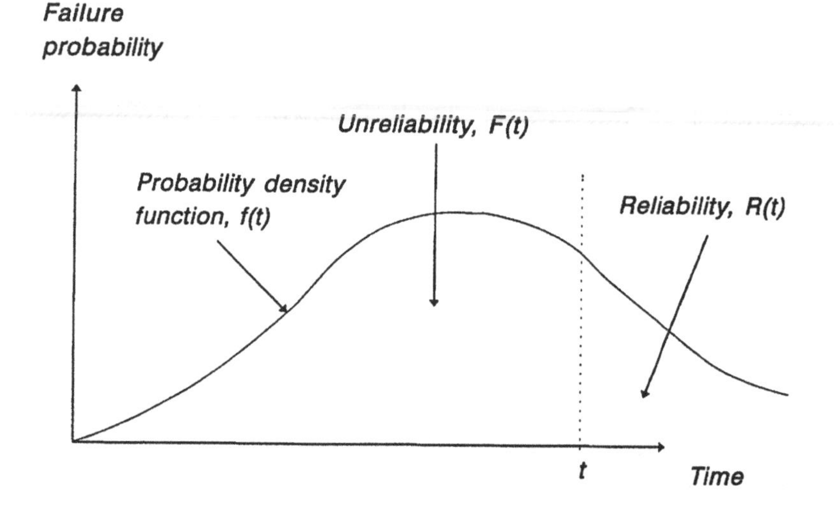

Figure 9.3 Relationship between f(t), F(t) and R(t).

The general forms of the curves for each of these functions is shown in Figure 9.4 which illustrates the three important stages in the reliability life-cycle, namely:

- **AB – early life failure**, sometimes referred to as 'infant mortality', where the product or system fails prematurely due normally to a design or raw material failure. This aspect of reliability is normally the focus for the first

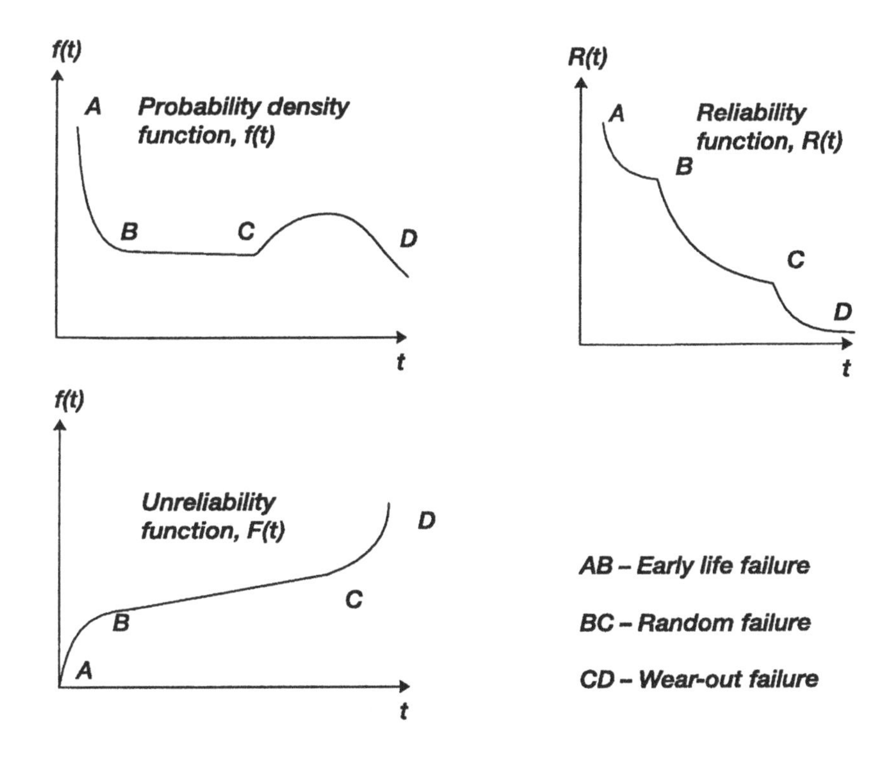

Figure 9.4 General forms of the failure characteristic functions.

stage of quality development and is reduced through the implementation of a quality systems approach.

- **BC – constant failure**, sometimes referred to as 'random failure', where the failure rate is approximately constant. This is often prescribed as the normal working life of the product or system and the reduction in the failure rate is generally brought about during the second stage of quality development where the focus is upon continuous improvement.
- **CD – wear-out failure**, sometimes referred to as 'old-age failure', which represents an increasing failure rate due to components or systems failing through incurring wear beyond the design or intended life. This final aspect of reliability is extended during the final stage of quality development during which there is an investment in a prevention orientation.

Another important characteristic function of reliability is the **age specific failure rate**, $\lambda(t)$. This is perhaps the most commonly known distribution as the shape of the curve is the classical 'bathtub' whereby the initially high failure rate falls to remain constant during the working life of the product only to increase again due to the cumulative ageing of the product, as shown in Figure 9.5.

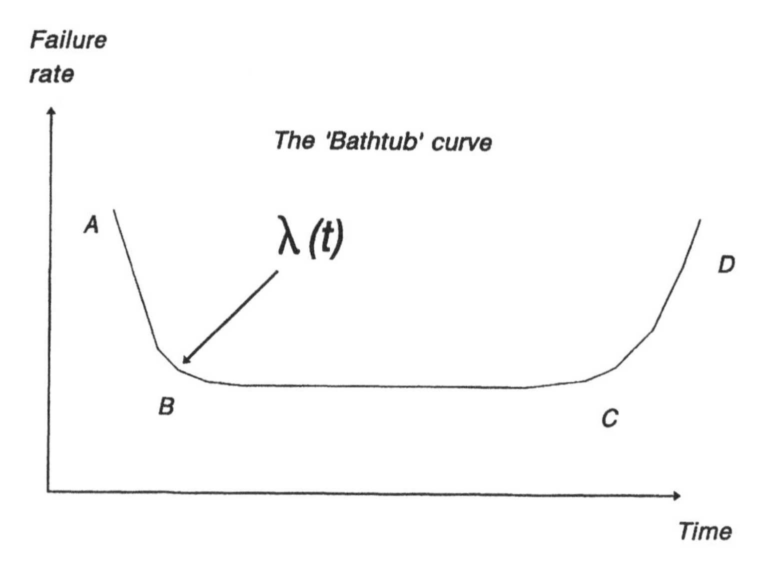

Figure 9.5 Age-specific failure rate life history curve.

In simple terms the failure rate is the ratio of the number of items failed to the total elapsed time (or number of operations):

$$\text{Failure rate } \lambda = \frac{\text{Number if items failed}}{\text{Total operating time (or cycles) of all items}}$$

The instantaneous value of this rate of failure is the age-specific failure rate $\lambda(t)$ and this may vary over time as illustrated by the bathtub curve.

This bathtub distribution intuitively mirrors the failure performance of many manufactured products. The motor car, for example, often exhibits unreliability when brand new as the vehicle undergoes 'teething' problems not detected at the pre-delivery inspection. A period of use for the motor car then exists (typically around 60 000 miles for modern cars) in which the reliability is fairly constant and subject only to the random failure of components. Finally the motor vehicle becomes unreliable as major steering, engine or transmission components begin to cumulatively fail.

When considering the maintenance or servicing of products or systems the two most commonly used characteristics of reliability are the **mean time to failure** (MTTF) or, where components are repaired, the **mean time between failures** (MTBF).

Again, in its most simple terms, the mean time to failure represents the average time to failure for all the items tested. Where the failure rate is constant then the MTTF is the ratio of the total elapsed time (or number of operations) to the number of items failing, in other words the inverse of the failure rate:

$$\text{MTTF} = \frac{1}{\lambda}$$

In terms of the variation of the failure probability with time (the probability density function, f(t)) a more comprehensive expression for the MTTF is given by:

$$\text{MTTF} = \int_0^\infty t.\text{f}(t).\text{d}t$$

Within maintenance management the concept of mean time between failures (MTBF) is more generally used as this is clearly a critical factor in determining the planned maintenance periodicity and the number of spare parts required. The problem in practice, however, in determining the correct value of MTBF is that the timing depends upon the nature of the repair which can be either:

- **maximized repair** whereby the component or system is effectively made as good as new and therefore assumes the failure rate of a new item; or
- **minimized repair** whereby the component or system is brought to the same condition in terms of failure rate as the item was in just prior to the failure.

Clearly if the simple constant failure rate model is adopted (as it is in many industrial preventative maintenance programmes) then the nature of the repair is not a significant factor and the MTBF is given by:

$$\text{MTBF} = \text{MTTF} = \frac{1}{\lambda}$$

where the failure rate, λ, is a constant.

Where the probability density function for a repaired unit $\text{f}_r(t)$ can be determined then the MTBF for the maintained component or system can be established from:

$$\text{MTBF} = \int_0^\infty t.\text{f}_r.\text{d}t$$

In servicing industries such as computer equipment repair or electric motor repair, very often there is sufficient data available to determine the failure characteristics of the repaired product and as a result warranty periods and spare part stocks are adjusted accordingly.

9.2.2 FAILURE DISTRIBUTIONS

In order to determine the useful reliability characteristics such as failure rate, MTTF and MTBF, there is therefore a need to be able to 'model' the failure probability with time, the probability density function, f(t).

From the appearance of the generalized curves shown in section 9.2.1 and one's own intuitive understanding that different items fail in different ways, it is clearly a great challenge in the management and analysis of reliability to obtain an appropriate expression for f(t).

A number of different formulae (or distributions) have been adopted to model the way in which items fail over time and the three most common types are:

- **exponential distribution** which is appropriate for modelling the failure characteristics where the failure rate is constant, i.e. region BC of the bathtub curve whereby failure is random;
- **normal distribution** which is appropriate in situations where the probability of failure is distributed about a mean value and the failure rate increases with time;
- **Weibull distribution** which uses a scaling constant which can be varied to reflect the reliability characteristic in each of the main regions of the bathtub curve, AB early life failure, BC random failure, CD wear-out failure.

The exponential distribution is probably the most widely used function to describe failure probability with time. This popularity is due to the combined benefits of simplicity and applicability. The probability density function is described using the exponential distribution as follows:

$$f(t) = \lambda . e^{-\lambda . t}$$

The failure function is therefore given by:

$$F(t) = \int_0^t f(t).dt = 1 - e^{-\lambda . t}$$

and hence the reliability function takes the simple form:

$$R(t) = 1 - F(t) = e^{-\lambda . t}$$

The assumption here is of course that the failure rate is constant and therefore is only applicable to the period during which failures are random, covered by the region BC on the bathtub curve. As this represents the normal working life of the product or system then the exponential distribution is applicable to the type of failure most commonly under consideration.

As stated above in section 9.2.1, the MTTF can be derived when the failure rate is constant as follows:

$$\text{MTTF} = \int_0^\infty t.f(t).dt = \int_0^\infty t.\lambda.e^{-\lambda . t}.dt = \frac{1}{\lambda}$$

An important corollary of this is seen when the reliability of the product or system is calculated at the MTTF:

$$R(\text{MTTF}) = e^{-\lambda . t} = e^{-\lambda . \frac{1}{\lambda}} = e^{-1} = 0.368$$

which means that only 36.8% of the items are expected to survive at the MTTF and therefore the maintenance period is normally set well within the MTTF.

The normal distribution is perhaps the least used of the functions although as expected it is an appropriate model for 'natural' components or systems. The normal distribution is appropriate where the failure of items is symmetrically distributed about the mean value and the failure rate increases with time. This model accurately depicts the normal wear-out failure of components or systems reaching old age or at the end of the design life. Neither premature failure nor random failure is appropriately described using the normal distribution.

The Weibull distribution is particularly important in the modelling of reliability due to the ability to describe each of the three main stages of the bathtub curve. The general form of the probability density function using the Weibull function is as follows:

$$f(t) = \frac{\beta}{\alpha} . [\frac{t - t_0}{\alpha}]^{\beta - 1} . e^{-[\frac{t - t_0}{\alpha}]^{\beta}}$$

and hence the failure function:

$$F(t) = 1 - e^{-[\frac{t - t_0}{\alpha}]^{\beta}}$$

and therefore the reliability function becomes:

$$R(t) = e^{-[\frac{t - t_0}{\alpha}]^{\beta}}$$

where: α = scaling constant;
β = shaping constant;
t_0 = locating constant.

The Weibull distribution is capable of modelling early life, random and wear-out failure through the selection of appropriate values for the shaping constant, β. The special cases are as follows:

β = 1 represents the region BC, random failure;
β < 1 represents the region AB, early life failure;
β > 1 represents the region CD, wear-out failure.

This flexibility of the Weibull distribution has a number of practical benefits in terms of its application to reliability management. For example, failure data on a new product can be evaluated using the Weibull distribution to determine the reliability performance of the product and indicate from the value of β whether the product is failing prematurely, randomly or due to ageing beyond the design life. Appropriate corrective actions in terms of the quality development tools and techniques to employ (as discussed in section 9.2.1 above) can then be specified based upon the value of β. Similarly if the value of β is known for a particular system, then the maintenance period can be determined using the Weibull distribution to ensure a specified level of system reliability. This flexibility has resulted in the growing popularity of the Weibull distribution as an approach to modelling reliability.

A comparison of the exponential, normal and Weibull probability density functions is shown in Figure 9.6. The corresponding failure rate curves are shown in Figure 9.7 together with the facility of the Weibull distribution to replicate the bathtub curve through the variation of the value of β.

9.2.3 RELIABILITY ANALYSIS

The simplest of the distributions used to describe the probability density function is the exponential function. The associated ease of use is, however, the main benefit in adopting this approach to the modelling of reliability.

As the failure rate is assumed constant using the exponential distribution then λ can be determined from experimental data by simply taking the ratio

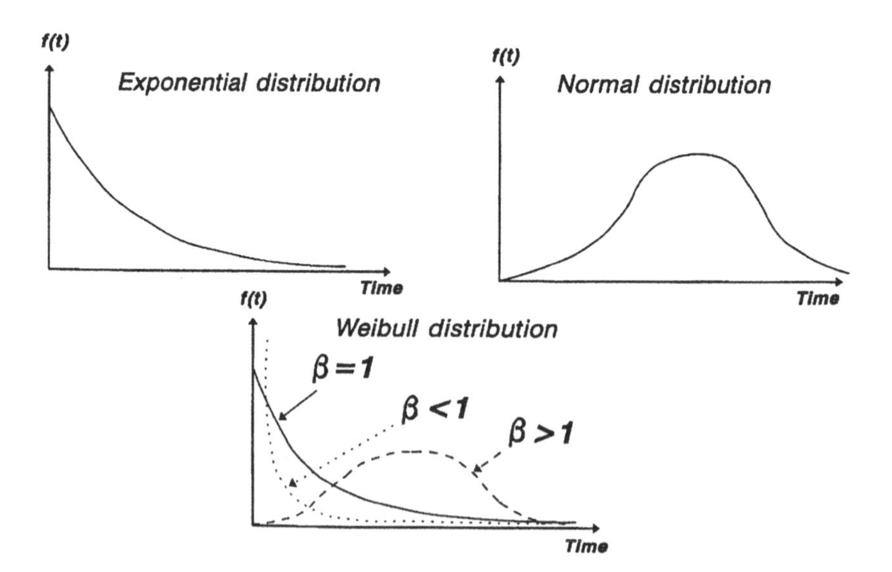

Figure 9.6 Frequency distributions for exponential, normal and Weibull functions.

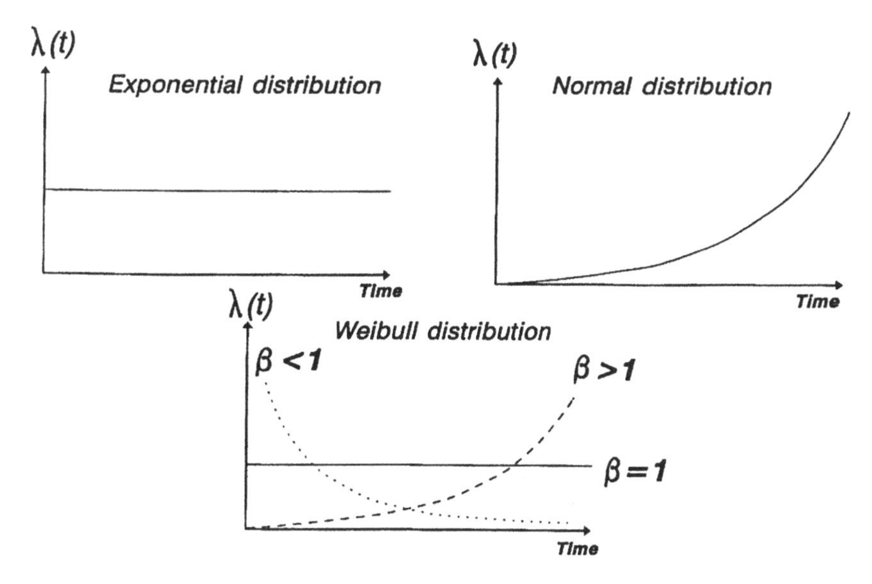

Figure 9.7 Failure rate as a function of time for exponential, normal and Weibull functions.

of the number of items failed to the total operating time (or number of cycles) as described above in section 9.2.1. With the failure rate known, the reliability of a component or system at a particular point in time can be determined as shown in the example below.

Example

Trial data shows 105 items fail during a test with a total operating time (for all failed and non-failed units) of one million hours. Hence:

$$\text{Failure rate, } \lambda = \frac{105}{10^6} = 1.05 \times 10^{-4}$$

What is the reliability of the item after 1000 hours?

$$\text{Reliability, R(1000)} = e^{-\lambda . t} = e^{-1.05 \times 10^{-4} \times 1000} = 0.9$$

Therefore an item has a 90% chance of surviving to 1000 hours.

As can be seen, this form of reliability analysis is relatively straightforward although the limitations emerge when the reliability for the next 1000 hours are considered. From section 9.3 on the reliability of components in series, the reliability of the item during the period 1000 hours to 2000 hours is determined from:

Reliability (0–2000 hours) = Reliability (0–1000 hours) × Reliability (1000–2000 hours)

Hence:

$$\text{Reliability (1000–2000 hours)} = \frac{\text{Reliability (0–2000 hours)}}{\text{Reliability (0–1000 hours)}}$$

$$R(1000-2000) = e^{-\lambda t_{1000-2000}} = \frac{e^{-\lambda \times 2000}}{e^{-\lambda \times 1000}} = \frac{0.81}{0.9} = 0.9$$

This states then that the reliability (probability of failure) for the second period of 1000 hours is exactly the same as for the first 1000 hours. This is contrary to the intuitive view that items become less reliable with time but the constancy of failure rate is of course fundamental to the assumption made in adopting the exponential distribution. Reliability analysis using the exponential model is therefore simple but limited.

The normal distribution again has the advantage of being easy to apply as the analysis of reliability requires reference to normal distribution tables such as shown in Appendix A. The normal distribution is appropriate for the analysis of reliability during the wear-out phase of the bathtub curve. The unreliability function is given by determining the area under the curve from the normal distribution table. The reliability is therefore the remaining area under the normal distribution curve as illustrated above in Figure 9.3.

Reliability analysis using the **Weibull** distribution involves the graphical determination of the Weibull parameters which can then be used in the expression for the reliability function. The graphical technique to determine the Weibull parameters makes use of special log-log × log paper (Chartwell graph paper ref.6572). The Chartwell system uses:

$$R(t) = e^{-[\frac{t-t_0}{\eta-t_0}]^{\beta}}$$

where η is the characteristic life and $\eta - t_0 = \alpha$ the Weibull scaling constant.

Using the log-log × log paper produces:

$$\ln[\ln(\frac{1}{R(t)})] = \beta \ln[\frac{t-t_0}{\eta - t_0}]$$

and therefore by plotting on the Chartwell paper the cumulative percentage failure against time (or cycles) to failure this will give a straight line (providing $t_0 = 0$) which can be used to estimate the shaping constant β. This constant effectively 'tunes' the Weibull distribution to model the appropriate phase of the bathtub curve. The construction used to determine the Weibull

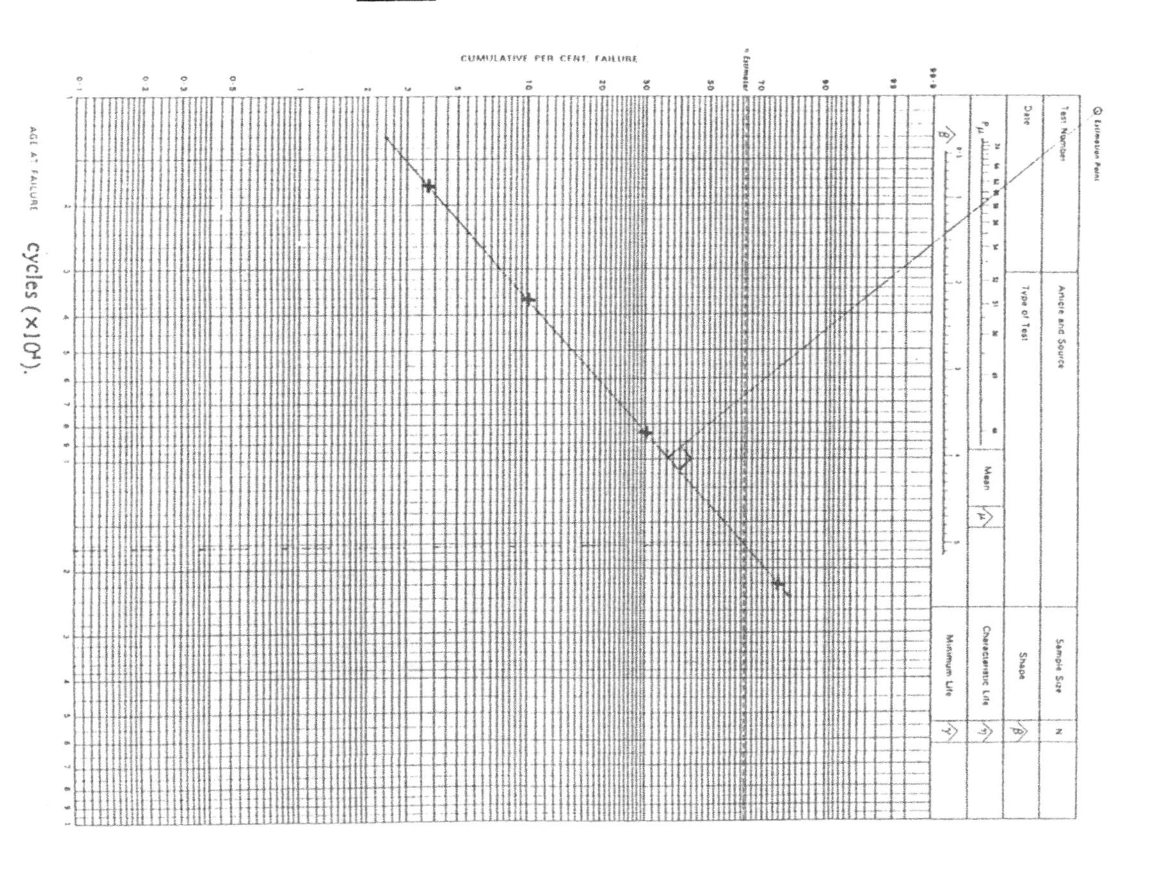

Figure 9.8 Example of Weibull plot to determine constants β and η.

From the Weibull plot: β = 1.45 ('old age' failure);

η = 17.5 × 10^4 cycles (characteristic life).

parameters is illustrated in Figure 9.8. The stages in the construction of the Weibull plot are as follows.

- **Stage 1.** Construct the graph from the experimental data showing cumulative percentage failure against time (or cycles) which should represent a straight line.
- **Stage 2.** Draw a line from the estimation point (printed towards the topleft-hand corner of the Chartwell paper) perpendicular to the straight line through the data.
- **Stage 3.** Read off from the β scale where the perpendicular constructed in Stage 2 crosses the scale. This scale gives directly the value of the shaping constant β.
- **Stage 4.** Using the dotted line shown on the Chartwell paper as the 'η estimator' read off at the point at which this estimator crosses the straight line through the data. This value of time (or cycles) is the value of η from which the scaling constant α can be determined (if $t_0 = 0$ then $\eta = \alpha$). The η estimator is established by taking the reliability at the value of $t = \eta$, hence:

$$R(t) = e^{-1} = 0.368$$

Therefore if 36.8% of the items survive then the characteristic life, η corresponds to 63.2% cumulative failure. The η estimator is printed onto the Chartwell paper at this level of cumulative failure.

Once the values of β and η have been determined the reliability at any point in time can be determined from the expression given above for the Chartwell system reliability function. In addition, the Chartwell paper can also be used to determine the mean life (MTTF) by taking the value of the percentage failure at the mean life (this is given by the point at which the perpendicular crosses this scale), then the percentage cumulative failure can be read off from the data giving the time at the mean life.

If, when the raw data is plotted onto the Chartwell paper, the plot is not a straight line then this indicates that the values for time to failure were not zeroed and therefore t_0 was not equal to zero. Typically such data exhibits a concave plot and the time to failure data needs to be corrected as follows.

1. Draw two sets of parallel lines equi-spaced in the cumulative failure plane.
2. Read off the corresponding values of time at failure t_1, t_2 and t_3 where the constructed lines cross the original data plot.
3. Calculate the correct value of t_0 using the expression

$$t_0 = \frac{t_3 t_1 - t_2^2}{t_3 + t_1 - 2t_2}$$

4. Correct the original data for the calculated value of t_0.

This construction is illustrated in Figure 9.9.

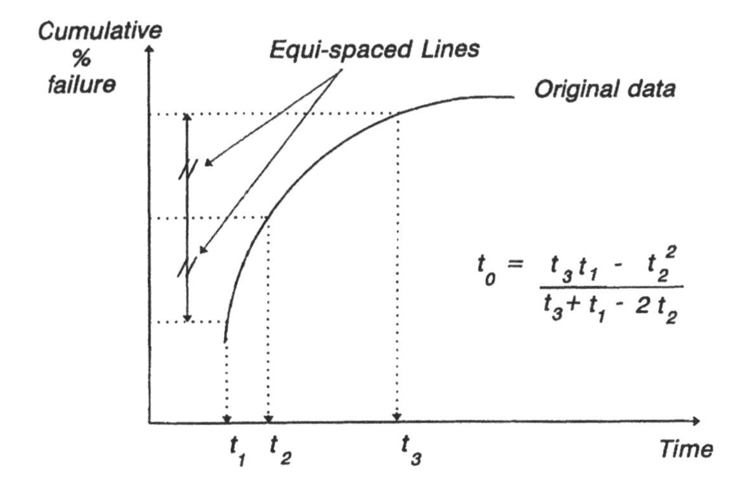

Figure 9.9 Construction for the correction of Weibull data when $t_0 \neq 0$.

In practice the interpretation of data on a Weibull plot is often extremely difficult due to the following:

- The failure of components is often not part of a 'controlled experiment' and may contain misrepresentative data caused by changes in the operating conditions.
- The information regarding the time at failure is often unreliable and in many instances the data represents the time at which the failure was detected or reported.

As a result the failure data from the launch of a new product in the marketplace or the operating time to failure for a new piece of process equipment is very often inaccurate and the resultant Weibull plot on the Chartwell paper is rarely exactly straight. Distinguishing between data which merely varies about a straight line mean and data which shows a true concave pattern requires some experience and informed analysis as to how the original data was collected.

Despite these difficulties, the Weibull distribution and graphical analysis technique is a very powerful reliability tool and an important contributor to quality development through improved product or process reliability.

9.2.4 SUMMARY

- The reliability of a product or system can be described in terms of its failure characteristics. These include the probability density function, f(t) the unreliability function F(t), the reliability function R(t), the failure rate $\lambda(t)$ and the mean time to failure or, for repaired items, the mean time between failures.

- The function describing the probability of failure with time can be simply modelled using either the exponential distribution, the normal distribution or the Weibull distribution.
- The reliability of a product or system can be determined using the exponential model by assuming a constant failure rate, from the statistical tables using the normal distribution and graphically using Weibull analysis.

9.3 The reliability of systems

9.3.1 THE APPLICATION AND CLASSIFICATION OF SYSTEMS RELIABILITY

The reliability analysis techniques described above in section 9.2 relate to the process of determining the probability of failure for a single entity which could be modelled using some form of failure distribution. In most practical situations, however, it is more common that the product or process under consideration comprises a number of separate entities combined to form a system.

To analyse and determine the reliability of components combined together to form systems requires the manipulation of probabilities. To analyse the combined reliability of a group of entities by using the individual probability density functions becomes extremely unwieldy even for the simplest of systems. The analysis of systems is normally undertaken using the reliability function, that is the probability of survival at a specified time.

The other basic assumption in the analysis of the reliability of systems is the concept of **independence**. It is assumed that the elements of a system fail independently and therefore the failure of any single item has no bearing upon the ongoing reliability of the surviving items. Clearly this assumption is inappropriate in situations where the failure of one item influences (often reduces) the reliability of other elements of the system. One approach to this limitation is to adopt a dual or multi-reliability model for the components of the system whereby the reliability probability can be amended (perhaps more than once) as other items within the system fail.

The formal analysis of systems comprising a number of components combined together requires the classification of certain system 'building blocks'. The basic system classifications for the analysis of reliability are as follows:

- **series** systems where the overall reliability requires all the components within the system to survive;
- **parallel** systems where the overall reliability requires any one of the components within the system to survive;
- **standby** systems where the system failure requires both the primary and standby elements to sequentially fail;

- **Bayes** systems in which components exist and influence the reliability of the overall system without being able to be classed as either series, parallel or standby.

The basic arrangement of each of these system building blocks is shown in Figure 9.10.

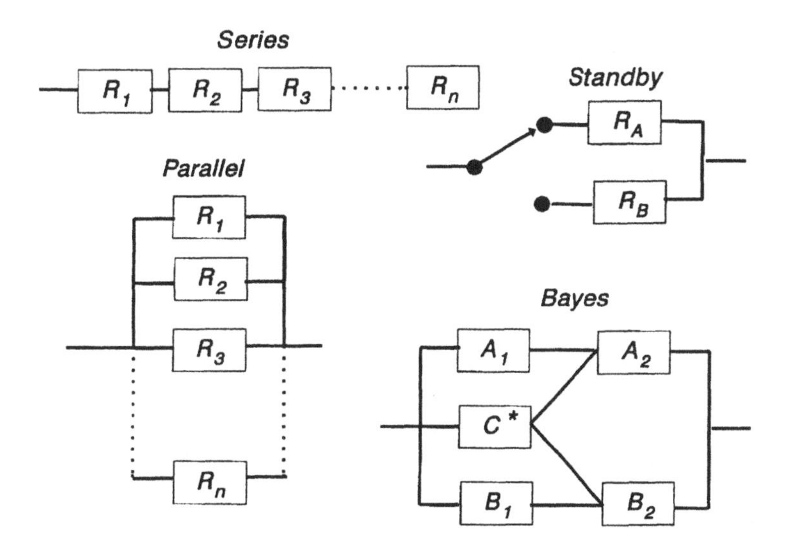

Figure 9.10 Systems and arrangement for series, parallel, standby and Bayes elements.

9.3.2 SYSTEMS RELIABILITY ANALYSIS

To analyse the effect of combining reliabilities it is important to understand two basic elements of probability theory:

- If A and B are two independent events with probabilities of occurrence of P(A) and P(B) then the probability that both events will occur is:

$$P(AB) = P(A).P(B)$$

In other words the system reliability is the product of the individual probabilities.

- If events A and B are complementary and mutually exclusive then:

$$P(A) + P(B) = 1$$

The reliability therefore of a system which comprises components arranged in series as shown above in Figure 9.10 is given by:

$$R(s) = R_1 \times R_2 \times R_3 \times \ldots \times R_n$$

For a system which requires all the component elements to survive, the reliability is the product of the individual reliabilities. As the individual reliabilities are generally less than 1 then the reliability of the system is less than the reliability of any individual component when configured in series.

When components are arranged in parallel to form a system then the system failure occurs only when all of the individual components fail. The failure of the system arranged in parallel is therefore given by:

$$F(s) = F_1 \times F_2 \times F_3 \times \ldots \times F_n$$

Now as a failure and survival are complementary and mutually exclusive then if the failure functions are replaced with one minus the reliability function, the expression for parallel systems becomes:

$$(1 - R(s)) = (1 - R_1) \times (1 - R_2) \times (1 - R_3) \times \ldots \times (1 - R_n)$$

or, where the components all have the same reliability R:

$$R(s) = 1 - (1 - R)^n$$

Unlike with components arranged in series, where components are arranged in parallel the overall systems reliability is generally greater than the reliability of the individual components.

In certain applications, however, it is neither possible nor desirable to improve the reliability of the system by arranging components in parallel. For example, the electrical characteristics of a system change when components are arranged in parallel or alternatively the energy consumption of a system may be increased if a second power source is operated in parallel. For these reasons use is often made of components arranged in standby in which a standby component is arranged in parallel but only brought into operation if the primary component fails, as illustrated above in Figure 9.10.

In such standby configurations, the probability that the system will survive is the probability of either component A surviving or component A failing and component B surviving. The system reliability for components arranged in standby is therefore given by:

$$\begin{aligned} R(s) &= R(A) + F(A) \times R(B) \\ &= R(A) + (1 - R(A)) \times R(B) \text{ (standby)} \end{aligned}$$

Most practical systems are a combination of series, parallel and standby arrangements as shown in Figure 9.11.

The approach to analysing such combined systems involves the reduction of series configurations to a single element and the reduction of any parallel or standby configurations to a single reliability. Through the process illustrated in Figure 9.11 even the most complex systems can be reduced to a simple series or parallel arrangement.

There are certain types of system configurations, however, which cannot

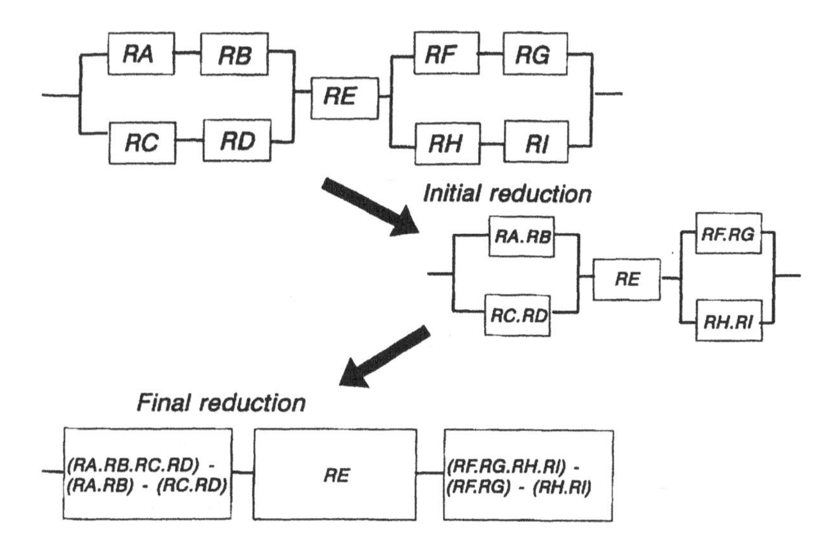

Figure 9.11 Combined systems reliability analysis.

be reduced to either series, parallel or standby systems. One of the most common examples of such a configuration is in systems in which a component has been installed in parallel with either of two other components as illustrated in Figure 9.12.

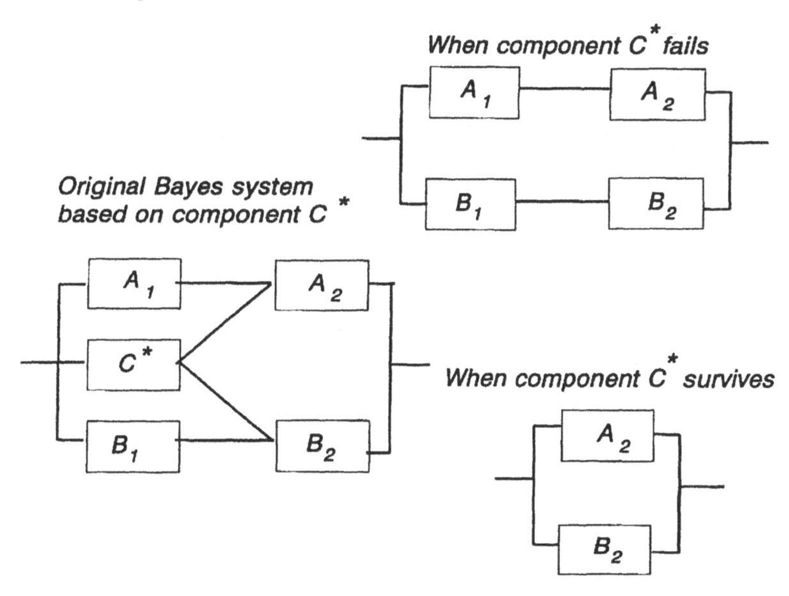

Figure 9.12 The Bayes component analysis.

This reasonably common approach to improving the system design reliability requires the use of Bayes' theorem for the analysis of the system including the Bayes component, C*. Bayes' theorem is applied by considering the reliability of the system when the Bayes component survives and also when the Bayes component fails. The failure probability for the system is then given by:

F(s) = F (Reduced system when the Bayes component survives) × (1 – F(C*)) + F (Reduced system when the Bayes component fails) × F(C*)

To analyse the failure probability of the reduced system, the configuration of the system needs to be determined in both of the cases, that is the Bayes component failing and the Bayes component surviving. This process is illustrated in Figure 9.12 which shows that if component C* survives then the reliability of the system does not depend upon the reliability of components A or B.

By substituting the reliability function in the above expression for the Bayes configuration, the system reliability becomes:

R(s) = R (system when C* survives) × R(C) + R (system when C* Fails) × (1 – R(C))

Much of the skill in the application of Bayes' theorem to the analysis of the reliability of systems is the identification of the Bayes component. A useful guide in the identification of the Bayes component is to examine for items which can contribute to the reliability of more than one sub-system.

9.3.3 SYSTEMS PREDICTABILITY AND AVAILABILITY

The reliability analysis techniques described above are of course concerned with the review of existing products or processes. The application of these techniques is very much part of the improvement-orientated organization's quality development efforts.

A prevention-orientated approach, however, demands that greater emphasis be given to the concept of reliability at the product or process design stage. To evaluate the potential reliability of a product or process at the design stage requires the application of predictive techniques, the two most common of which are:

- failure mode and effects analysis;
- fault tree analysis.

Failure mode and effects analysis (FMEA) is described in more detail in Chapter 7 in the context of statistical process control (SPC). Process FMEA is primarily used as a pinpointing technique within SPC to determine the critical characteristics of the process as described in section 7.2.

In terms of the application of FMEA to reliability management, the design FMEA is used to assess potential failures of the product due to component or sub-system unreliability. The design FMEA is normally undertaken by a team selected from the marketing, production, sales and service functions in addition to the design team. The objective is to review the product design and to establish the following.

- What are the main functional elements of the design?
- What are the potential failures?
- What are the effects of the failure?
- What are the main causes of failure?
- What are the controls used to prevent failure?

From this analysis an assessment is made of the failure potential of the design and this is expressed as the risk priority number (RPN). The reliability of the component or the sub-system (as determined using the exponential, normal or Weibull distributions) is used to calculate the 'occurrence' value. Where the value of the RPN is high (typically greater than 90) this indicates the need for design corrective action to prevent the likelihood of reliability problems.

In terms of the process FMEA this can be used by maintenance managers to identify processes with high failure risks and therefore to focus maintenance corrective actions and preventative maintenance programmes.

The second predictive technique, **fault tree analysis (FTA)**, is a valuable method for understanding the relationships between elements within a system and how a particular mode of failure depends upon other failure events. The value in undertaking FTA is that it can often assist in the prediction of major failures due to the failure of minor components or sub-systems. Detection mechanisms can be implemented for the minor failures to prevent the occurrence of the generally more expensive major failure.

Such detection mechanisms are more apparent once the system has been systematically analysed using FTA. The process of generating an FTA involves determining the logical connectivity between events using 'and' and 'or' type logic. For example, the failure of an engine system can be predicted from the failure of an elemental component such as the lubrication sub-system as shown in Figure 9.13.

The predicted reliability performance of a product or process is generally measured in terms of the **availability**. This measure is particularly important in the maintenance of systems and is usually expressed as either the operational availability (A_O) which is expressed in terms of the mean time between maintenance (MTBM) and the mean downtime (MDT) as given by:

$$A_o = \frac{\text{MTBM}}{\text{MTBM} + \text{MDT}}$$

or alternatively for the design of processes or products, the inherent availability (A_I) is used in terms of the mean time between failure (MTBF) and

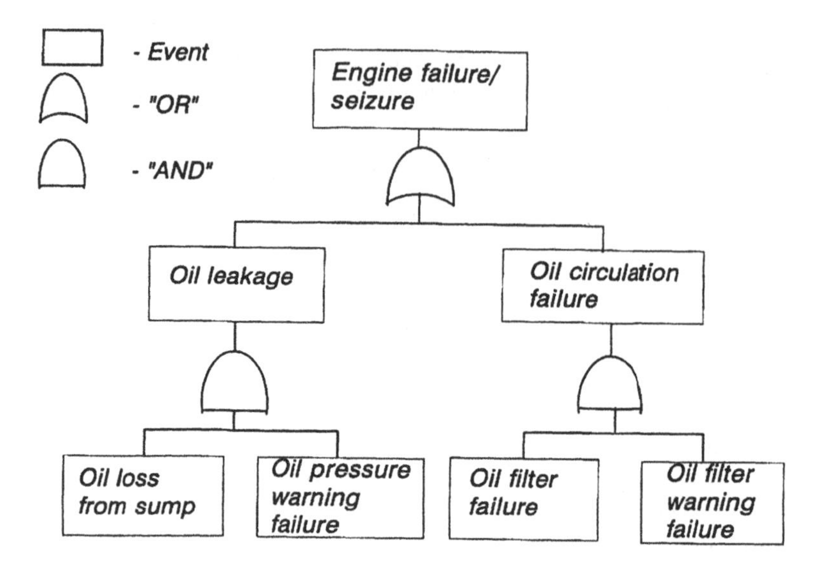

Figure 9.13 Fault tree analysis.

the mean time to repair (MTTR) as given by:

$$A_I = \frac{\text{MTBF}}{\text{MTBF} + \text{MTTR}}$$

The inherent availability does not assume any preventative maintenance downtime (as this cannot be predicted at the design stage) and is therefore a design prediction of the compromise relationship between reliability and maintainability.

9.3.4 SUMMARY

- Systems reliability can be determined by considering the arrangement of series, parallel, standby or Bayes sub-systems.
- Reliability analysis involves the manipulation of probabilities and combined systems can be simplified by reduction to enable the overall reliability to be determined.
- The predicted reliability of products or processes can be improved using FMEA or FTA and the availability measured in terms of either operational or inherent availability.

10 Advanced quality planning

10.1 The prevention approach to quality management

10.1.1 THE ROLE OF QUALITY ENGINEERING IN QUALITY DEVELOPMENT

As organizations mature in the way in which quality is managed, so the emphasis changes from the control of the process to the design. Designing products and processes in which quality problems are prevented rather than detected is not only more cost effective, but also produces more consistent products and services. The application of quality management tools and techniques to the design of products or processes is generally termed **advanced quality planning** or **quality engineering**.

The emphasis in quality engineering is upon designing **robust** products and processes which are tolerant to inherent variabilities within the manufacturing process and the application of the product or service. These variations to which the design needs to be tolerant are often termed **noise**. There are three basic forms of noise to which the product or service may be subjected:

- outer noise (such as variations in environmental conditions);
- inner noise (variation due to deterioration or wear within the product);
- between product noise (variations from piece to piece due to imperfections in the manufacturing process).

The robustness of the design is therefore the stability of the product to remain within specification when subject to the inherent noise factors. The **signal to noise ratio** is a term that is used to describe specifications and the noise factors. Robust design is achieved through a prevention-orientated approach in which both products and processes are designed to be tolerant to the noise variation expected.

The application of advanced quality planning techniques is part of the third stage of quality development (as described in Chapter 1) in which these prevention-orientated techniques dominate. In order to apply these techniques an organization needs to have clearly defined customer requirements, a clear understanding of material and process variability, an established culture of teamworking and a passion for improvement. Such organizations are therefore already mature in terms of their quality development and this prevention orientation represents an advanced stage of development.

10.1.2 THE ELEMENTS OF ADVANCED QUALITY PLANNING

The prevention-orientated approach to quality management is one in which the emphasis is put upon the product and process design activities rather than product inspection or process control activities. This distinction between **on-line** quality activities (such as acceptance sampling or statistical process control) and **off-line** quality activities (such as quality function deployment or design of experiments) reflects the changing emphasis during the quality development of an organization. Rather than evaluating the product or service after it has been produced, the initial stage of development focuses instead upon controlling the production process and finally the emphasis changes to the design activities with the techniques of advanced quality planning.

One of the key contributors to the techniques of advanced quality planning was Dr Genichi Taguchi. In terms of the role of quality management in the design process, Taguchi identified three principal aspects of design:

- system design;
- parameter design;
- tolerance design.

The **system design** relates to the technical knowledge and background necessary to specify the complete functional design. This process involves relating the customers' requirements to product functionality, and the key technique for providing system design is quality function deployment as described below in section 10.2.

Parameter design is used to establish the process parameters which need to be used to reduce the sensitivity of the design to noise. The key prevention-orientated technique for establishing the optimized settings for process parameters is design of experiments as described below in section 10.4. The objective of parameter design is to identify process settings which produce robustness to noise factors and increases the signal to noise ratio.

Tolerance design optimizes the allowable range for the product or process settings by identifying the balance between the costs of conformances and the costs of non-conformance. Traditionally, establishing the tolerances in the design of products or processes is done by some form of technical convention and does not attempt to optimize the settings. Quality loss function is the primary advanced quality planning technique for establishing tolerance design.

The relationship between system, parameter and tolerance design is illustrated in Figure 10.1.

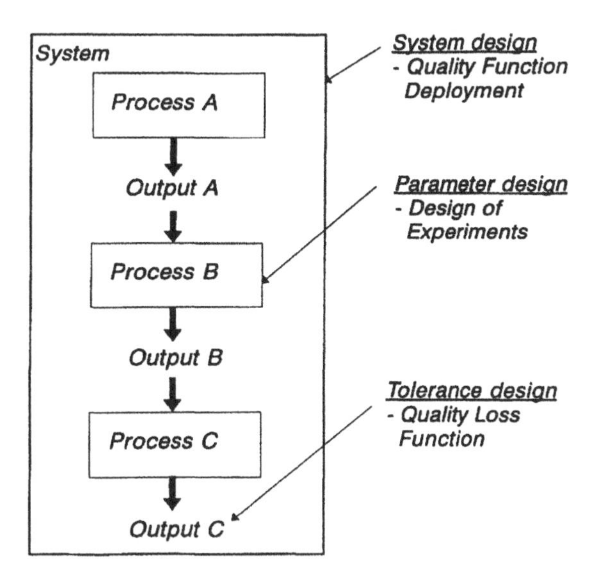

Figure 10.1 System, parameter and tolerance design.

10.1.3 THE BENEFITS OF ADVANCED QUALITY PLANNING

The application of the techniques of advanced quality planning represent an investment in prevention which in turn reduces the overall cost of quality (as described in Chapter 3). The cost benefits from the application of quality management techniques to the design of the product or service arise from:

- reduced costs of non-conformances due to more robust designs which are more in line with the customer's requirements and more tolerant of noise factors;
- reduced need for modifications to the product or process specifications which are generally increasingly costly as the product or service moves from the design stage to the manufacturing stage to the application by the customer stage.

The diminishing returns from devoting increasing organizational effort to inspection and verification activities encourages companies to become improvement orientated, to reduce process variation and to solve the basic problems which cause defects. Eventually improvements in products and processes which are not robustly designed also exhibit diminishing returns and the organization needs to place more emphasis upon the initial design of the product or process by adopting a prevention-orientated approach through the application of advanced quality management tools. The evolution of this process in terms of the changing emphasis in the way quality development is managed is illustrated by the experiences of Japanese manufacturers shown in Figure 10.2.

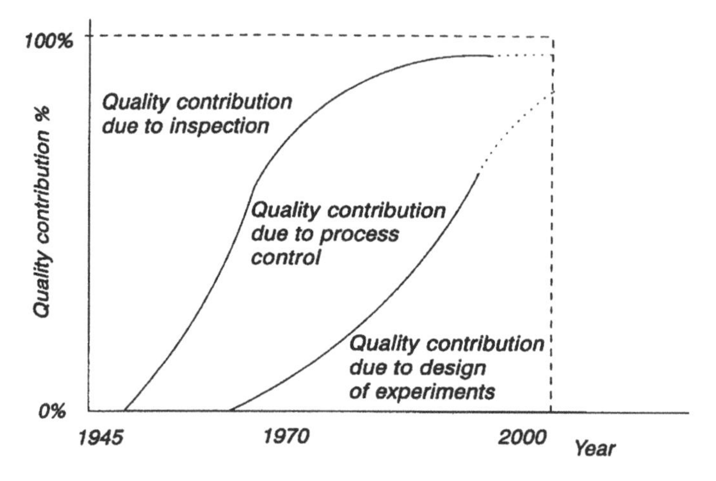

Figure 10.2 The changing contribution to quality development of quality management techniques in Japan.

Once manufacturing variability is minimized and the quality system is effectively implemented then the major area of quality development opportunity available for most organizations is improved product and process design. Typically for mature companies, where the processes are in control and the systems and procedures well established, then the quality of the product or service is determined largely by the design process (rather than the manufacturing process). Many manufacturers of electronic equipment or motor cars now believe that the quality of the product is defined in the design of both the product and the manufacturing process.

The benefits associated with better designs which require fewer subsequent modifications is illustrated in the comparison between Japanese and US manufacturers during the 1980s shown in Figure 10.3.

Having to make changes to a new product (for example, a new soap powder) after the product has been marketed and launched is not only technically expensive (in terms of the changes to process specifications) but is even more costly in terms of the impact in the marketplace.

10.1.4 SUMMARY

- As organizations develop in the way in which quality is managed then greater emphasis is placed upon product or process design to improve the robustness to noise.

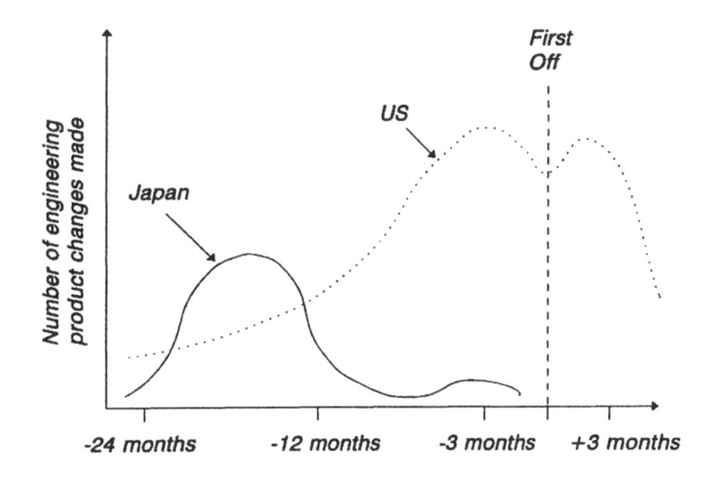

Figure 10.3 Comparison of engineering design changes between Japanese and US companies.

- The techniques of advanced quality planning assist in system design (quality function deployment), parameter design (design of experiments) and tolerance design (quality loss function).
- The benefits of advanced quality planning including the reduction in both the cost of quality and the number of subsequent modifications to product or process specifications has increased the emphasis upon design quality in mature, prevention-orientated organizations.

10.2 System design and quality function deployment

10.2.1 SYSTEM DESIGN AND CUSTOMER REQUIREMENTS

One of the most demanding organizational quality challenges is to design products or services which actually meet the requirements of the customer. The problem often encountered is that the requirements of the customer are very often expressed in a format and a language very different from the product or process specifications used by the designer. For example, the customer-important characteristics of a biscuit (texture, appearance, taste, etc.) are very often not quite the same as the product specification (weight, moisture content, colour). Similarly the customer's requirement from a life insurance policy (reliability, ease of claiming for next of kin, etc.) are expressed in different terms to the specification of the insurance policy (actuarial rates, maturity benefits, etc.). The abstraction needed from a system designer to convert customer requirements (often expressed in subjective, qualitative terms) into product performance (usually defined objectively in quantitative terms) is considerable. The primary quality management technique to assist in this process and make the translation of customer requirements more systematic is **quality function deployment (QFD)**.

The basic objective of QFD is to systematically relate the requirements of the customer (termed the 'voice of the customer') to the features of the product (termed the 'counterpart characteristics'). QFD relates these characteristics through a set of matrices which build to form the 'house of quality' so called because of the structure of the way in which the relationships are represented. Having established the relationship between the product characteristics and the customer's requirements, the QFD technique then evaluates the relative importance of the requirements, compares performance with (potential) competitors' performance and then deploys the important requirements down to the sub-components and to the manufacturing processes. Overall QFD generates four different diagrams or matrices as illustrated in Figure 10.4.

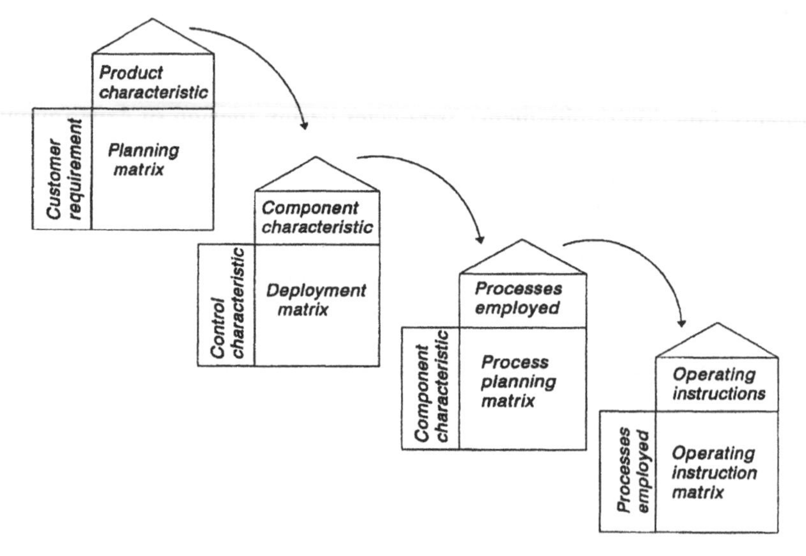

Figure 10.4 The four matrices documented in quality function deployment.

The first of these matrices is termed the **planning matrix** which relates the voice of the customer to the specific counterpart characteristics.

The second of the matrices is the **deployment matrix** which decomposes the product characteristic requirements down to the requirement of the major components. This is a mechanism for ensuring the customer's requirements are deployed down to the characteristics of the components of the product and are reflected in the component design.

The third of the matrices is termed the **process planning and quality control chart**. It identifies the critical control points which ensure that the product and sub-component characteristics are controlled during the manufacturing process. Control points represent those aspects of the process which need to be checked to ensure the correlations between the product or component characteristics and the customer requirements are maintained.

Finally the **operating instructions matrix** further deploys the critical product and component parameters down to the operating instructions used by process operators.

This overall approach of integrating customer requirements to the product characteristics process controls and operating procedures represents a rigorous methodology for product systems design.

10.2.2 THE STAGES TOWARDS QUALITY FUNCTION DEPLOYMENT

Of the four 'houses of quality' associated with the QFD approach (and described above in section 10.2.2) it is the planning matrix which is most commonly used as this is the technique which relates customer needs to product features. The other three matrices/charts of QFD are, however, also important to the quality planning process as they deploy the relationships identified in the planning matrix down through the organization by defining the controls and procedures necessary.

The creation of the planning matrix involves eight stages as described below in Table 10.1. Stage 1 of the process in many ways is the most exacting QFD activity as the customer requirements are often difficult to establish in terms of what the customer expects from the product or service and why the customer purchases the product or service. Not surprisingly therefore the

Table 10.1 Creation of a QFD planning matrix

Stage 1	Definition of customer requirements.
Stage 2	Identify the product (counterpart) characteristics.
Stage 3	Develop the relationships between the requirements (Stage 1) and the characteristics (Stage 2).
Stage 4	Add the customer priorities and market evaluations to the requirements dimension of the matrix.
Stage 5	Add the competitive evaluations of the control characteristics.
Stage 6	Identify the selling points of the new design in terms of the requirements.
Stage 7	Develop target values for the control characteristics.
Stage 8	Select the control characteristics which need to be deployed down to the sub-component level.

customer requirements information is provided by a number of sources: market research, customer interviews, sales/marketing department, service reports, etc. This aspect of QFD is very much easier for companies which have developed a strong customer orientation and have close relationships with their customers as described in Chapter 5. The customer requirements are usually divided into primary, secondary and tertiary needs to further refine the definition of requirements. In some cases problems in identifying customer requirements actually point to a lack of understanding as to who the customer actually is!

For example, if a planning matrix is developed for the provision of an undergraduate lecture course in quality management the question arises 'who is the customer?' The most obvious answer would appear to be the undergraduate student; however, a more fundamental viewpoint is that the true customer is society (who after all meets most of the costs). The student's requirements for personal and professional development are similar but not the same as society's needs for education, classification and stimulation of the people leaving higher education. Similarly a manufacturing company may have intermediate customers (for example, retailers) whose requirements are different to those of the end users.

Stage 2 of the process identifies the counterpart characteristics which are those technical features of the product or service which enable the customer requirements to be met. The counterpart or product characteristics should be expressed in measurable terms and any relationships between the characteristics should be noted in the 'roof' of the house. These relationships are graded very strong, strong or weak according to the impact of one characteristic on another. So, for example, in the provision of an undergraduate course on quality management, there would be a clear correlation between the number of lecture hours and the breadth of techniques covered. This relationship matrix is built up to improve the understanding of the interdependency between characteristics and to appreciate how a change in one characteristic may affect another.

Stage 3 of the development of the planning matrix involves identifying the relationships between the customer requirements and the product characteristics. The significance of the relationship is again rated as either very strong, strong or weak and the matrix diagram is completed as shown in Figure 10.5. The importance of this stage in the process is that by formally identifying the relationships between requirements and characteristics any omissions can be readily identified. If, for example, a customer requirement did not have a very strong relationship with any of the product characteristics then it is reasonable to conclude that the requirement is not being met. Similarly a redundant characteristic may be identified as not having a relevance to any of the customer's requirements.

Stage 4 brings the customer's priorities to the 'garage' of the matrix together with a competitive evaluation of how the product or service

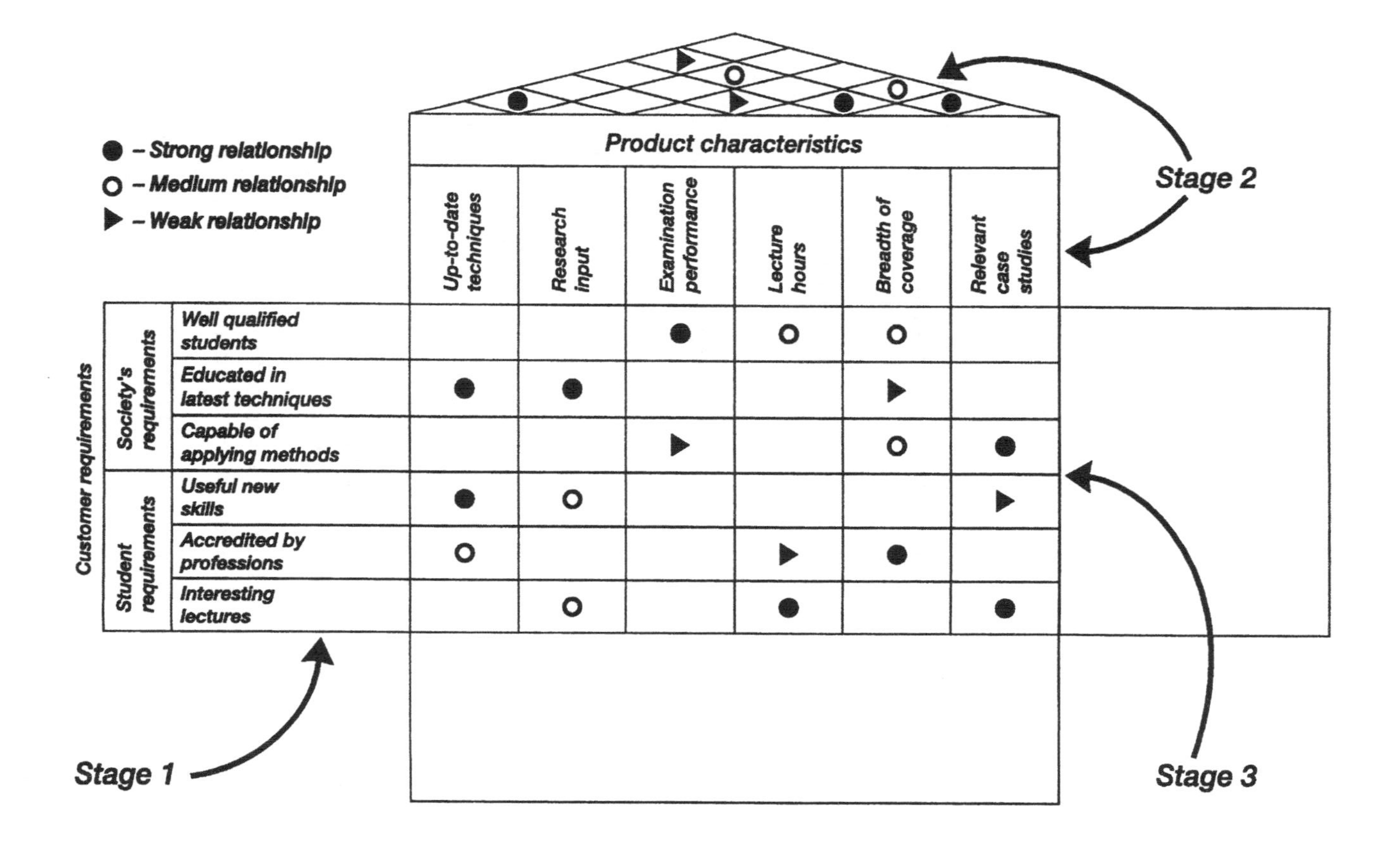

Figure 10.5 Stages 1 to 3 of the QFD planning matrix.

compares with that of competitors in terms of meeting the requirements of the customer. The prioritizing of requirements in terms of customer expressed importance is normally established through some form of market research and care must be taken here to identify any changing needs or priorities. The competitive evaluation data reflects the strengths and weaknesses of the new product in comparison to those of competitors and when these are considered in conjunction with the customer priorities they offer an important insight into how well the product will be received in the marketplace.

Stage 5 undertakes a competitive comparison similar to the evaluation undertaken in Stage 4, except the comparison is made in terms of the product characteristics rather than the customer requirements. The data for comparing the product characteristics with competitors' performance is normally obtained from a set of tests or evaluations stored in the 'foundations' and is expressed in objective, measured values. Consistency needs to be checked between the market-based competitive evaluations produced in Stage 4 and the results of the comparison of product characteristics generated in Stage 5. If, for example, the customer evaluation indicates a comparative disadvantage over a competitor's product yet the comparison of product characteristics shows an advantage this would imply some form of inconsistency in the methods of evaluation or that the relationship between the requirements and the characteristics is weaker than anticipated.

Stage 6 lists the key selling points which may then be advertised as part of the marketing of the new product. These selling points can be targeted at those requirements of the customer which are designated as most important in Stage 4. The QFD process therefore ensures that the key features of the new product which are to be used to market and promote the product are deployed down through the organization to operational controls and procedures.

Stage 7 develops target values for the product characteristics based upon the agreed selling points, the ranking in terms of customer importance and the strengths and weaknesses of the product. The targets established should be measurable values which can be evaluated in terms of the performance of the finished product. These target values contribute to the robustness of the design.

Stage 8 selects those characteristics which need to be deployed down to the remaining QFD process and charts. The selection of which characteristics need to be deployed again depends upon the key selling points identified, the customer priorities, the competitive evaluations and technical difficulties involved in achieving the target values.

The completed planning matrix is illustrated in Figure 10.6.

The creation of the deployment matrix involves basically two additional stages as described in Table 10.2. Stage 9 lists both the customer requirements and the product characteristics in greater levels of details for the critical elements deployed from Stage 8. The relationships between the

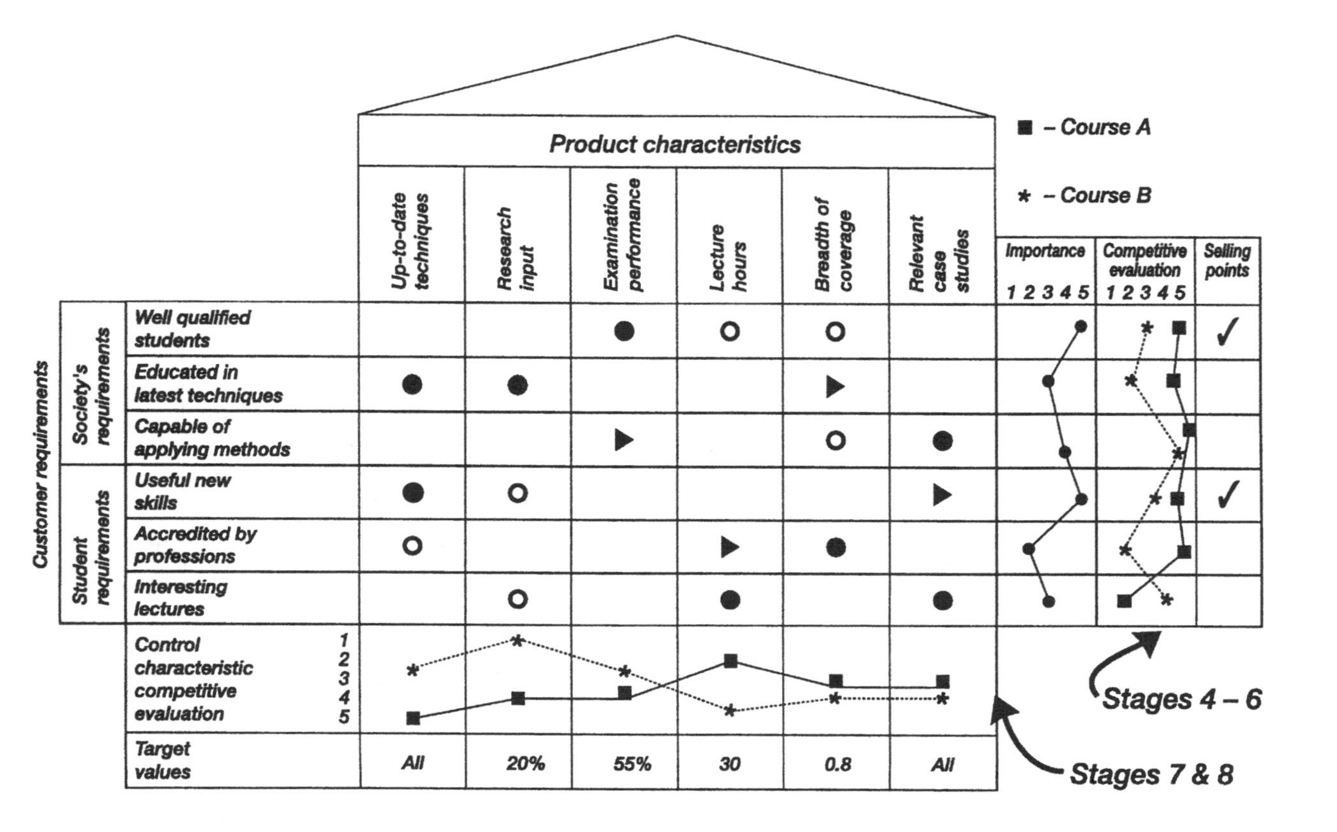

Figure 10.6 Stages 4 to 8 of the QFD planning matrix.

Table 10.2 Creation of a QFD deployment matrix

Stage 9	Deploy the QFD process down to the sub-component level in terms of both requirements and characteristics.
Stage 10	Develop the component deployment chart relating critical component control characteristics to critical sub-component control characteristics.

characteristics of the sub-component and the finished product characteristics are identified at this stage. These relationships are then used to identify which of the sub-component characteristics need to be controlled.

Stage 10 takes the deployment further by defining the sub-component control characteristics needed to achieve the target values for the finished product characteristics. These relationships are depicted on the component deployment chart which lists for each of the sub-components the correlation between the finished product characteristic and the corresponding critical sub-component characteristic.

An example of a deployment matrix is shown in Figure 10.7.

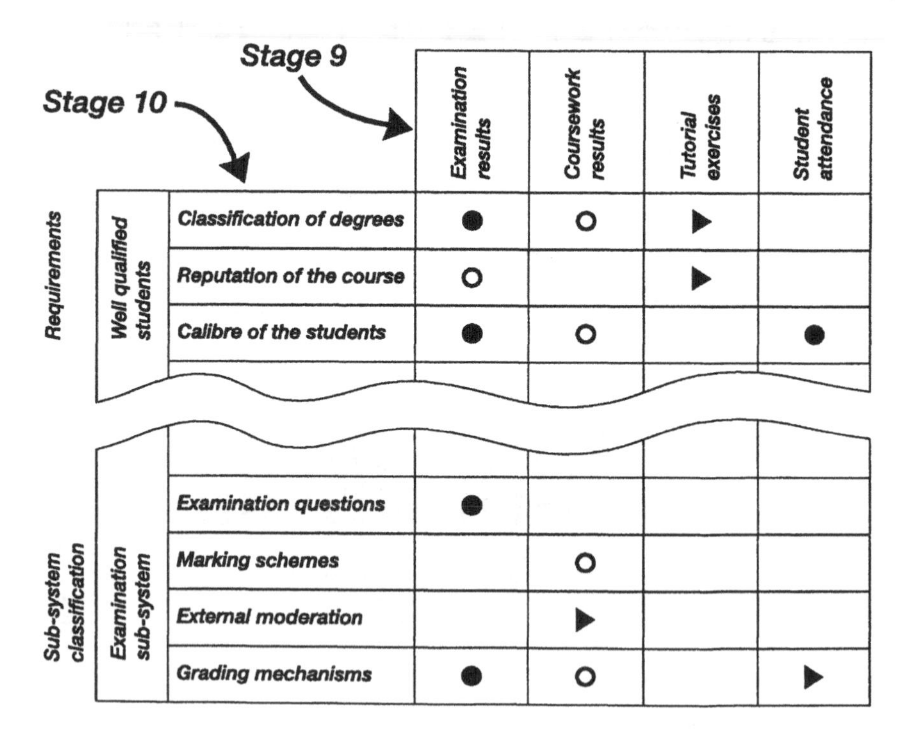

Figure 10.7 The QFD deployment matrix.

The next phase of the QFD process creates the process plan and control charts. This comprises two stages as described in Table 10.3.

Table 10.3 Creation of a QFD process plan and control chart

Stage 11	Develop the relationships between the critical characteristics and the processes used to create the characteristics.
Stage 12	Develop the control plan relating critical controls to critical processes.

Stage 11 represents the transition from the design of the product to the production of the product. The preparation of the process plan includes the identification of 'control points' which are those process steps which directly affect a critical product (or sub-component) characteristic. These critical control points are then the basis for establishing the quality control activities such as acceptance sampling (Chapter 6) or statistical process control (Chapter 7).

Stage 12 expands the activities associated with the control points identified in Stage 11 by identifying the control plan parameters such as the sample size/frequency and the method of checking, and also describes the process flow.

The activities associated with Stages 11 and 12 are similar to the stages involved in the preparation of a process failure mode and effects analysis (FMEA) diagram as described in Chapter 7. A design FMEA is very much an equivalent technique to the preparation of a process control plan using QFD. The process plan and control chart produced by Stages 11 and 12 is illustrated in Figure 10.8.

Processes	*Control points*	*Control methods*	*Process controls*
Set examination questions	*- Question difficulty* *- Question relevance*	*- External examiner*	*External regulation*
Define marking scheme	*- University guidelines*	*- Notes for guidance*	*Annual comparisons*
Mark examination papers	*- Allocation of marks*	- $\bar{X}$ *and* σ	*Cross-marking of scripts*

Figure 10.8 The QFD process planning and control chart.

The final phase of the QFD process is the preparation of the operating instructions. This is a single stage as described in Table 10.4.

Table 10.4 Creation of a QFD operating instructions

Stage 13	Tabulate operating instructions from process requirements, check points and control points.

Stage 13 basically defines the process methods which should be employed to ensure that the process requirements are met and the controls and checks required at the critical points in the process to ensure the control plan is carried out. The format for operating instruction sheets tend to be company-specific and can vary in the level of detail necessary to effectively define the process.

In total, the 13 stages of QFD represent an integrated quality planning tool for ensuring the requirements of the customer are reflected throughout the organization's quality activities.

10.2.3 IMPLEMENTING QUALITY FUNCTION DEPLOYMENT

Although the preparation of the planning matrix (Stages 1 to 8 in section 10.2.2) is often considered the essence of QFD, all four of the 'houses of quality' are important contributors to the advanced quality planning process. The nature of QFD requires the process to be a team-based activity with contributions from marketing, design, technical, production and inspection personnel. Effective teamwork is therefore an organizational requisite for the implementation of QFD.

The rigorous, step-by-step nature of QFD needs to be maintained during implementation rather than an attempt made to carry out some of the activities in parallel. To realize the benefits of this systematic approach, many organizations implementing QFD validate the process rigour by 'signing off' the matrices and charts at each stage of the process. The operation of QFD therefore becomes both a project management mechanism for the planning of a new product and a design review mechanism. The QFD process described in the 13 stages of section 10.2.2 is intended as a framework for implementation and individual organizations can tailor the approach to suit the particular application in terms of the nature of the product or service, the format of the marketing data and the complexity of the operational processes.

QFD has made a significant contribution to the design and development of more quality orientated new products or services. Companies such as Toyota have seen a reduction in the product development cycle times together with fewer engineering changes required as a result of implementing QFD. The QFD process does, however, require a more prevention-orientated approach and an investment in effort prior to the manufacturing process. The implementation of QFD also represents a powerful technique for developing customer orientation throughout the organization.

10.2.4 SUMMARY

- Quality function deployment is an important advanced quality planning technique for ensuring the requirements of the customer are reflected in the characteristics of the product or service.
- The QFD process comprises the building of four 'houses of quality': the planning matrix, the deployment matrix, the process plan and control chart and the operating instruction sheet.
- The implementation of QFD is a team-based activity which promotes a prevention-orientated approach to quality management.

10.3 The quality loss function

10.3.1 THE DEFINITION OF QUALITY LOSS FUNCTION

The approach to quality engineering promoted by Taguchi is based upon quality of a product being considered as: 'the (minimum) loss imparted by the product to the society from the time the product is shipped'. From this definition Taguchi represented the quality loss as a quadratic function – that is the loss varies with the square of the deviation of the quality characteristic, as illustrated in Figure 10.9.

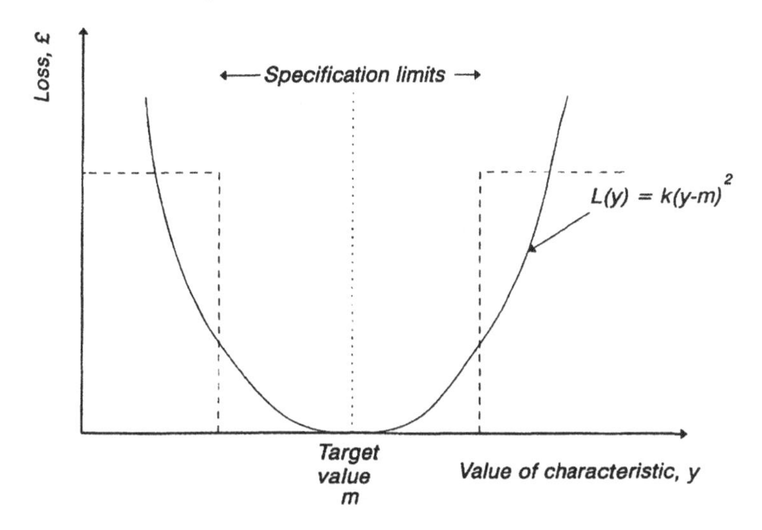

Figure 10.9 The quality loss function.

The quality loss function is therefore given by:

$$L(y) = k(y - m)^2$$

where: $L(y)$ = the monetary loss per unit of product when the quality characteristic is equal to the value y;
y = the value of the characteristic (for example, length, concentration, voltage, etc.);
m = the target value of the characteristic y;
k = the proportionality constant which depends upon the financial significance of the product to society.

This expression for loss implies a continuous function whereby the loss increases as soon as the value of the characteristic varies from the target. This approach is to be contrasted with the traditional view of quality in which the cost of non-conformance is only incurred once the product characteristic falls outside the specification limits as discussed Chapter 3.

The features resulting from the quality loss function being considered as a quadratic expression are:

- the loss is minimized when $y = m$;
- the loss initially increases slowly but increases more rapidly as the value of y moves away from the target m;
- the loss is expressed in financial terms.

By adopting this approach to quality engineering, whereby the objective is to produce characteristics which meet the target rather than simply fall within specifications, a more robust design is produced. This is particularly true for the design of complex products in which the cumulative effect of components being away from target value can have a significant impact upon the overall reliability of the product.

The quality loss function also enables tolerance design to be undertaken from the viewpoint of the overall loss to society rather than tolerances being selected simply on the basis of material or process capability.

10.3.2 QUALITY CHARACTERISTICS

The first stage in quality engineering involves identifying the characteristics of the product or service which are to be measured. Taguchi classifies these quality characteristics as either:

- **measurable (variable) characteristics** which are definable according to some continuous scale;
- **attribute characteristics** which are measured according to some form of acceptable/unacceptable criteria which are often subjective; or
- **dynamic characteristics** which are the properties of a 'system' and are determined on the basis of the relationship between the input to the system and the resulting output.

The measurable characteristics can be further classified in terms of the target value for the quality loss function. Three classifications are used:

- **nominal-the-best** – in which the characteristic has a specific target value (for example, dimensions, purity or density);
- **smaller-the-better** – where the target for the characteristic is ultimately zero (for example, wear, deterioration or impurities);
- **larger-the-better** – where the target is ultimately infinity (for example, strength, useful life or efficiencies).

For the nominal-the-best characteristic the quality loss function is minimized when the characteristic y approaches the target value m and is given by:

$$L(y) = k(y - m)^2 \quad \text{Nominal-the-best}$$

as described above in section 10.3.1.

For the smaller-the-better characteristic the quality loss function is minimized when the characteristic y approaches zero and is given by:

$$L(y) = ky^2 \quad \text{Smaller-the-better}$$

Finally, for the larger-the-better characteristic the quality loss function is minimized when the characteristic y approaches infinity and is given by:

$$L(y) = k(\frac{1}{y})^2 \quad \text{Larger-the-better}$$

A comparison of the three types of measurable characteristic is shown in Figure 10.10.

The quality loss functions described above do, however, relate to the loss associated with only one unit. In most practical cases more than one unit is produced and for the nominal-the-best characteristic the quality loss function becomes:

$$\begin{aligned} L(y) &= k\frac{(y_1 - m)^2 + (y_2 - m)^2 + (y_3 - m)^2 + \ldots + (y_n - m)^2}{n} \\ &= k[\frac{1}{n}\sum_{i=1}^{n}(y_i - m)^2] \\ &= k[[\frac{1}{n}\sum_{i=1}^{n}(y_i - \bar{y})^2] + (\bar{y} - m)^2] \\ &= k[\sigma^2 + (\bar{y} - m)^2] \end{aligned}$$

where: $\sigma^2 = \frac{1}{n}\sum_{i=1}^{n}(y_i - \bar{y})^2$

The σ^2 term describes the **consistency** of the batch and the $(y - m)^2$ term is a measure of the **accuracy**. These concepts of variability and position are described in more detail in Chapter 7.

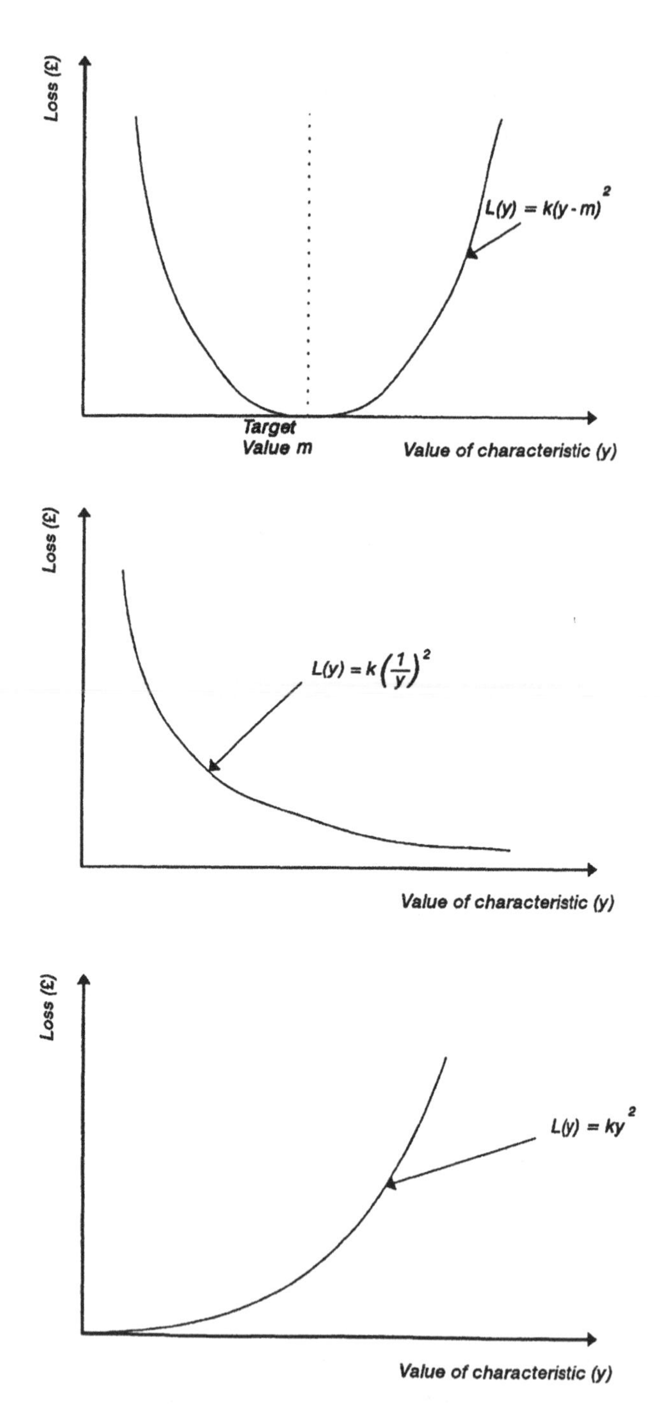

Figure 10.10 A comparison of the three types of measurable characteristics.

The loss associated with a particular process can therefore be reduced by either improving the consistency (reduced variability) or by improving the accuracy (adjusting the mean to the target).

Similarly the quality loss function for more than one unit can be expressed for the smaller-the-better characteristic as:

$$L(y) = \sigma^2 + \bar{y}^2$$

and for the larger-the-better characteristic as:

$$L(y) = \frac{1}{n}\sum_{i=1}^{n}\frac{1}{y_i^2}$$

10.3.3 TOLERANCE DESIGN

One of the important applications of the quality loss function is in the design of manufacturing tolerances. The relationship between the characteristic y and the loss $L(y)$ can be used to establish the point at which the loss to society (the combined producer's loss and consumer's loss) is minimized.

Consider, for example, the setting of the tracking angle for the front wheels of a car which has a target value of –1.5 degrees. If the cost of resetting the tracking angle at the end of the production line is £8.00, what should the manufacturing tolerance be? From a quality loss function perspective the design of the tolerances requires an evaluation of the tracking angle at which the manufacturer justifiably spends £8.00 correcting the vehicle. Assuming the cost of rectifying a tracking problem (which occurs at an angle of +1°) for a customer who returns the new vehicle to the dealer under warranty is £140.00 (including the cost of worn tyres!) then the manufacturing tolerance can be calculated as shown in Figure 10.11.

From Figure 10.11:

$$L(y) = k(y - m)^2$$

$$\text{for } y = 1^\circ,\ L = £140,\ m = -1.5^\circ$$

$$\text{hence } k = \frac{140}{(2.5)^2} = 22.4\ £/^\circ$$

The manufacturing tolerance can therefore be found by calculating the value of tracking angle (y) which corresponds to a loss of £8.00. Hence:

$$L(y) = £8 = k(y - m)^2$$

$$\therefore \pm\sqrt{\frac{8}{22.4}} + m = y$$

$$\text{hence } y = -1.5^\circ \pm 0.6^\circ$$

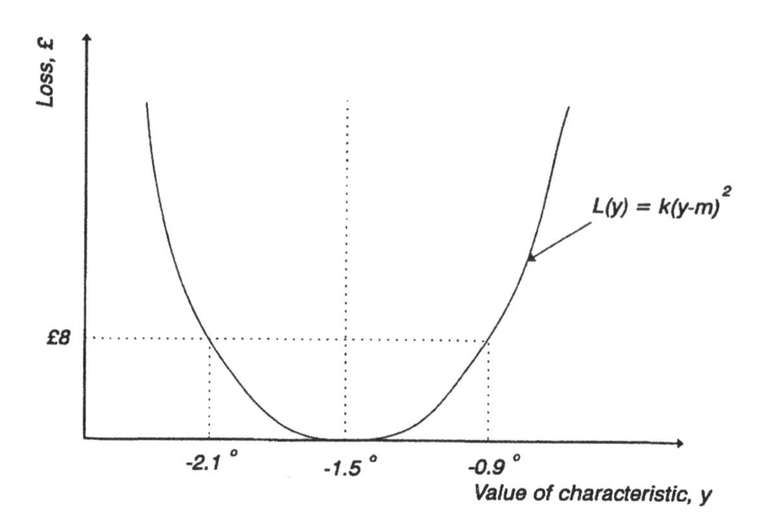

Figure 10.11 Tolerance design using the quality loss function.

The manufacturing tolerance is therefore plus or minus 0.6 degrees. If the vehicle is manufactured with the tracking outside these limits then the manufacturer should reset the tracking angle otherwise the loss to society is not optimized.

This approach to tolerance design attempts to integrate both the manufacturer's and the customer's quality cost requirements.

10.3.4 SUMMARY

- The Taguchi quality loss function defines the monetary loss to society as a quadratic expression based upon the variation from the target value.
- The measurable quality characteristics can be classified as either nominal-the-best, smaller-the-better or larger-the-better.
- Manufacturing tolerances can be designed based upon the quality loss function to optimize the loss to society.

10.4 Design of experiments

10.4.1 FACTORIAL DESIGN OF PROCESSES

The third element of quality engineering is parameter design in which the process settings are established to ensure that the manufacture of the product or service is optimized. By determining the parameters which optimize the signal (accuracy) to noise (consistency) ratio then more robust products or services are produced and fewer non-conformances will reach the customer. In simple terms the signal to noise ratio is the vulnerability of the performance of the product or service to the disturbances to which it is likely to be subjected.

The design of process parameters therefore needs to consider both types of factors, namely:

- **control factors** which are those process parameters whose level can be set and maintained;
- **noise factors** which are those parameters whose level cannot be set or controlled but affect the performance of the characteristic.

By examining the relationships between control factors and noise factors a more robust design of process settings can be established. The essence of parameter design is to conduct a set of experiments to determine the optimum setting for the process parameters.

The techniques for the design of experiments are not new; indeed, some of the classical methods were developed by R.A. Fisher in the UK in the early part of this century. Much of this early experimental design work was used in the optimization of yield in farming. Taguchi developed an approach to applying the techniques of experimental design to industrial processes and whilst the simplified approach adopted by Taguchi has been criticized by statisticians, it does represent a useful and practical technique for optimizing process parameters.

The simplest form of process experiment is to consider one parameter at a time. For example, if a manufacturing process has seven factors (temperature settings, speed settings, raw material purity, etc.), each of which, for simplicity, can be set at one of two levels (high/low, fast/slow, high purity/low purity) then a process experiment could be carried out in which one factor at a time is varied and the process yield (efficiency) measured as shown in Figure 10.12.

For each experiment the 'best' setting for the factor is determined and this is then selected and the next factor evaluated. For a process having seven factors a total of only eight experiments are required as shown in Figure 10.12. Whilst this appears to be an efficient approach to optimizing the process settings, there is a major assumption in this technique that the factors do not interact. If in the above example there is an interaction between factors B and E, say, the 'one factor at a time' approach does not fully evaluate the process (the result for B or E is not measured with both factors in either of their two states, i.e. the B1 with E2 result, for example, is not obtained).

Many industrial processes are difficult to control and to establish robust parameter settings for precisely because of the interactions present. The 'optimum' settings established using a 'one factor at a time' experiment may not in fact represent the most robust combination of process settings.

	Factor							
Experiment number	*A*	*B*	*C*	*D*	*E*	*F*	*G*	*Experimental result*
1	*A1*	*B1*	*C1*	*D1*	*E1*	*F1*	*G1*	**Result 1**
2	*A2*	*B1*	*C1*	*D1*	*E1*	*F1*	*G1*	**Result 2**
3	*A2*	*B2*	*C1*	*D1*	*E1*	*F1*	*G1*	**Result 3**
4	*A2*	*B2*	*C2*	*D1*	*E1*	*F1*	*G1*	**Result 4**
5	*A2*	*B2*	*C2*	*D2*	*E1*	*F1*	*G1*	**Result 5**
6	*A2*	*B2*	*C2*	*D2*	*E2*	*F1*	*G1*	**Result 6**
7	*A2*	*B2*	*C2*	*D2*	*E2*	*F2*	*G1*	**Result 7**
8	*A2*	*B2*	*C2*	*D2*	*E2*	*F2*	*G2*	**Result 8**

Comparison A1 vs A2, result 2 is 'best'
etc. for 8 experiments

Figure 10.12 'One factor at a time' experiment.

An alternative approach would be to undertake a full factorial experiment where each of the seven factors is evaluated at both levels. This approach to parameter design would, however, require 27 combinations – in other words 128 process experiments – to be carried out to determine the optimum process settings. In most practical situations it would be unrealistic to conduct such a large number of experiments and if the process factors have three levels rather than two, then the number of experiments required escalates to over 2000.

The approach to the design of experiments proposed by Taguchi involves the use of **orthogonal arrays** to reduce the number of experiments necessary to establish the optimum process parameters. An experimental array is described as orthogonal if the pairs of columns are balanced as illustrated in Figure 10.13).

Considering the first two factors in the array, A and B, it can be seen from Figure 10.13 that the number of occurrences of each of the factors at each of the levels is the same:

$$\text{For } A_1{:}\ \frac{\text{Number of experiments at } B_1}{\text{Number of experiments at } B_2} = \frac{1}{1}$$

similarly:

$$\text{For } A_2{:}\ \frac{\text{Number of experiments at } B_1}{\text{Number of experiments at } B_2} = \frac{1}{1}$$

Experiment number	Factor							Experimental result
	A	B	C	D	E	F	G	
1	1	1	1	1	1	1	1	y1
2	1	1	1	2	2	2	2	y2
3	1	2	2	1	1	2	2	y3
4	1	2	2	2	2	1	1	y4
5	2	1	2	1	2	1	2	y5
6	2	1	2	2	1	2	1	y6
7	2	2	1	1	2	2	1	y7
8	2	2	1	2	1	1	2	y8

$L_8(2^7)$ Orthogonal array
7 Factors
2 Levels each
8 Experiments

Figure 10.13 Orthogonal array for seven factors (L_8).

All other pairs of columns in the L_8 array can also be seen to be orthogonal. The orthogonal array represents a significantly reduced experimental design and enables the selection of optimal parameters without the large number of experiments needed to test the full range of factorial dependency.

10.4.2 PROCESS OPTIMIZATION USING ORTHOGONAL ARRAYS

In order to select an appropriate orthogonal array the process needs to be evaluated for the number of **degrees of freedom**. The number of degrees of freedom for all orthogonal arrays is one less than the number of experiments required. For the L_8 array described above in section 10.4.1 the total degrees of freedom is 7 (one less than the number of experiments, 8).

In terms of the process factors a degree of freedom is a measure of the amount of information that can be obtained from experiment. The degrees of freedom for factor A (process temperature setting, say) is the number of comparisons that need to be made for the various levels (settings) of factor A. So, for example, if factor A has two levels of setting (high/low) then factor A has one degree of freedom as only one comparison needs to be made. A degree of freedom is also associated with the interactions between factors. If factor A exhibits an interaction with a second factor B (for example, the process temperature interacts with, say, the power setting for the process) then again a number of comparisons (degrees of freedom) are required to obtain sufficient experimental information. The number of degrees of freedom for an interaction is equal to the product of the number of degrees of freedom for the individual factors. So, for example, if both factors A and B both had two possible levels (settings) then each factor would have one degree of freedom and the interaction would have $1 \times 1 = 1$ degree of freedom also. Hence for a process that has two process parameters, A and B, each of which can have two levels, and given that A and B interact, then the overall process has 3 degrees of freedom and therefore requires an orthogonal array which contains $3 + 1 = 4$ experiments. This is provided by the L_4 orthogonal array as shown in Figure 10.14.

In most practical situations, the degree of interaction between process parameters can be anticipated at the process design stage. If, however, the degree of interaction is not known (for example, in an entirely new process) then if lines are constructed between the values of the process characteristic at level A1 and A2 for both the B1 setting and the B2 setting then the degree to which the lines are parallel is a measure of interaction. If the lines cross then there is a strong interaction between the factors A and B as shown in Figure 10.14. The third column in the L_4 array is therefore for the 'factor' which is the interaction between A and B. The two 'levels' of interaction represented in the array are in fact a measure of whether the two basic factors A and B are in phase (both high or both low, for example) or out of phase (one high and one low).

Having identified the appropriate process factors to evaluate, including interactions, and selected the corresponding orthogonal array, the optimiza-

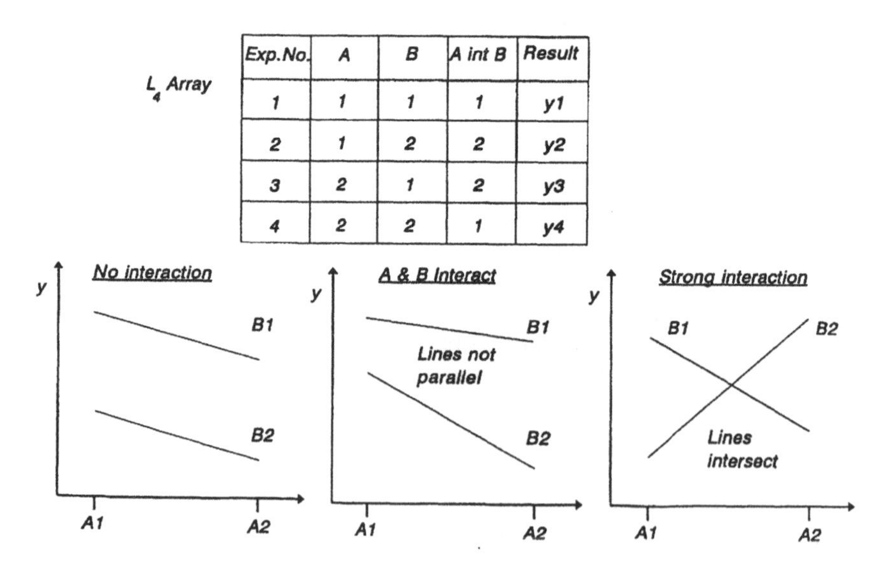

L_4 Array

Exp.No.	A	B	A int B	Result
1	1	1	1	y1
2	1	2	2	y2
3	2	1	2	y3
4	2	2	1	y4

Figure 10.14 Degrees of freedom for the L_4 orthogonal array.

tion process is relatively straightforward. From conducting the experiments prescribed in the array a set of results in terms of the process characteristics to be optimized (for example, yield, efficiency or strength) are obtained. For each of the factors in turn the average value of the characteristic is calculated for each level of the factor. So, for example, for the L_4 array shown in Figure 10.14 the setting of factor A is optimized by comparing the values of y at level A1 with the values of y at level A2:

$$\text{A1 average} = \frac{y1 + y2}{2} \quad \text{B1 average} = \frac{y1 + y3}{2}$$

$$\text{A2 average} = \frac{y3 + y4}{3} \quad \text{B1 average} = \frac{y2 + y4}{2}$$

The level for each of the factors which produces the highest average characteristic is then selected as the optimum process design. If the difference between the level 1 and level 2 average is significant then the factor is considered more important to set at the optimum level. Where the difference is small the factor is less important to optimize.

10.4.3 IMPLEMENTING DESIGN OF EXPERIMENTS

The techniques for the design of experiments are an important approach to developing more robust processes and are associated with the preventative stage to managing quality.

In their simplest application experimental orthogonal arrays can be used to identify the optimum settings for a range of process factors. This represents an important and practical quality development technique as many industrial processes have a number of factors, the setting of which is an important contributor to product quality. The results of a typical experimental design for a manufacturing process having seven factors (and seven degrees of freedom) is shown in Figure 10.15. By calculating the average mill efficiency for each of the factors the optimum setting can be calculated as A1, B2, C2, D2, E2, F1, G2.

The difference between the average efficiency for each factor obtained at level '1' and level '2' is a measure of how important it is that the particular factor should be optimized. Where the difference between the averages is large (for example, with factor B, compactability) then this is a significant factor in the setting of the process. For other factors, however, the difference is small (for example, factor D, permeability) and therefore the choice of level 1 or 2 is not significant and may be optimized for some other criteria, such as process cost.

In most practical applications of parameter design experiments, both control factors and noise factors need to be evaluated. The layouts used to evaluate both control and noise factors utilize inner (control) and outer (noise) arrays as illustrated in Figure 10.16.

A Moisture	B Compact	C % G.S.	D Permeability	E Shatter	F Available bond	G Working bond	H Mill efficiency
1	1	1	1	1	1	1	57.98
1	1	1	2	2	2	2	58.50
1	2	2	1	1	2	2	63.54
1	2	2	2	2	1	1	65.57
2	1	2	1	2	1	2	58.56
2	1	2	2	1	2	1	54.22
2	2	1	1	2	2	1	58.42
2	2	1	2	1	1	2	60.72

Figure 10.15 Example of the application of experimental design.

	L_8 inner array							L_4 outer array				
	Control factors							Noise factors				
Exp. No.	A 1	B 2	C 3	D 4	E 5	F 6	G 7	1	2	2	1	S/N ratio
								1	2	1	2	
								1	1	2	2	
1	1	1	1	1	1	1	1					
2	1	1	1	2	2	2	2					
3	1	2	2	1	1	2	2					
4	1	2	2	2	2	1	1					
5	2	1	2	1	2	1	2					
6	2	1	2	2	1	2	1					
7	2	2	1	1	2	2	1					
8	2	2	1	2	1	1	2					

Figure 10.16 The L_8 inner array for control factors and the L_4 outer array for noise factors.

The signal to noise ratio for the **nominal-the-best** process characteristic can be calculated from:

$$\text{Signal-to-noise ratio (dB)} = 10 \log \left[\frac{\bar{y}^2}{v_s} - \frac{1}{n}\right]$$

$$\text{where } v_s = \text{sample variance} = \sum_{i=1}^{n} \frac{(y_i - \bar{y})^2}{n-1}$$

The application of orthogonal arrays to parameter design requires a structured approach to initially identify the characteristics, to plan the experimental design and finally to evaluate the results and select the optimum settings which create a process which is robust to noise. The basic stages in the implementation of the design of experiments is shown in Figure 10.17 together with a comparison with the 'traditional' approach to process optimization.

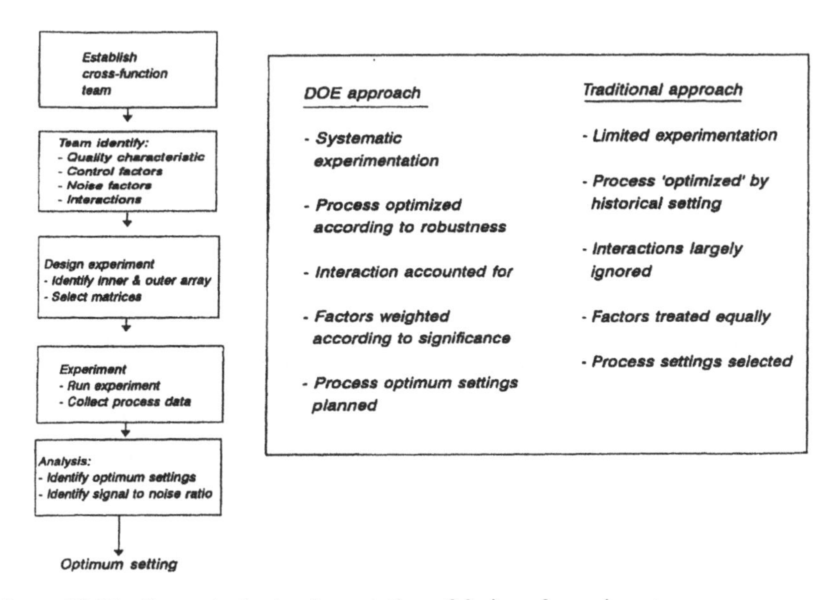

Figure 10.17 Stages in the implementation of design of experiments.

10.4.4 SUMMARY

- Parameter design involves process experimentation to establish the optimum settings of factors which influence the robustness of the process.
- Orthogonal arrays can be used to reduce the extent of the experimentation and these are evaluated using the average yield and signal to noise ratios.
- The design of experiments represents a systematic approach to process optimization.

Appendix A Area under the normal distribution curve

Tabulated values represent the area under the curve from ± α to the value of $X_i - \mu/\sigma$ where X_i is any value of the variable X, μ is the mean value and σ is the standard deviation of the population.

Table A1 Areas under the normal curve

$\frac{X_i - \mu}{\sigma}$	0.09	0.08	0.07	0.06	0.05	0.04	0.03	0.02	0.01	0.00
−3.5	0.00017	0.00017	0.00018	0.00019	0.00019	0.00020	0.00021	0.00022	0.00022	0.00023
−3.4	0.00024	0.00025	0.00026	0.00027	0.00028	0.00029	0.00030	0.00031	0.00033	0.00034
−3.3	0.00035	0.00036	0.00038	0.00039	0.00040	0.00042	0.00043	0.00045	0.00047	0.00048
−3.2	0.00050	0.00052	0.00054	0.00056	0.00058	0.00060	0.00062	0.00064	0.00066	0.00069
−3.1	0.00071	0.00074	0.00076	0.00079	0.00082	0.00085	0.00087	0.00090	0.00094	0.00097
−3.0	0.00100	0.00104	0.00107	0.00111	0.00114	0.00118	0.00122	0.00126	0.00131	0.00135
−2.9	0.0014	0.0014	0.0015	0.0015	0.0016	0.0016	0.0017	0.0017	0.0018	0.0019
−2.8	0.0019	0.0020	0.0021	0.0021	0.0022	0.0023	0.0023	0.0024	0.0025	0.0026
−2.7	0.0026	0.0027	0.0028	0.0029	0.0030	0.0031	0.0032	0.0033	0.0034	0.0035
−2.6	0.0036	0.0037	0.0038	0.0039	0.0040	0.0041	0.0043	0.0044	0.0045	0.0047
−2.5	0.0048	0.0049	0.0051	0.0052	0.0054	0.0055	0.0057	0.0059	0.0060	0.0062
−2.4	0.0064	0.0066	0.0068	0.0069	0.0071	0.0073	0.0075	0.0078	0.0080	0.0082
−2.3	0.0084	0.0087	0.0089	0.0091	0.0094	0.0096	0.0099	0.0102	0.0104	0.0107
−2.2	0.0110	0.0113	0.0116	0.0119	0.0122	0.0125	0.0129	0.0132	0.0136	0.0139
−2.1	0.0143	0.0146	0.0150	0.0154	0.0158	0.0162	0.0166	0.0170	0.0174	0.0179
−2.0	0.0183	0.0188	0.0192	0.0197	0.0202	0.0207	0.0212	0.0217	0.0222	0.0228
−1.9	0.0233	0.0239	0.0244	0.0250	0.0256	0.0262	0.0268	0.0274	0.0281	0.0287
−1.8	0.0294	0.0301	0.0307	0.0314	0.0322	0.0329	0.0336	0.0344	0.0351	0.0359
−1.7	0.0367	0.0375	0.0384	0.0392	0.0401	0.0498	0.0418	0.0427	0.0436	0.0446
−1.6	0.0455	0.0465	0.0475	0.0485	0.0495	0.0505	0.0516	0.0526	0.0537	0.0548
−1.5	0.0559	0.0571	0.0582	0.0594	0.0606	0.0618	0.0630	0.0643	0.0655	0.0668
−1.4	0.0681	0.0694	0.0708	0.0721	0.0735	0.0749	0.0764	0.0778	0.0793	0.0808
−1.3	0.0823	0.0838	0.0853	0.0869	0.0885	0.0901	0.0918	0.0934	0.0951	0.0968
−1.2	0.0895	0.1003	0.1020	0.1038	0.1075	0.1057	0.1093	0.1112	0.1131	0.1151
−1.1	0.1170	0.1190	0.1210	0.1230	0.1251	0.1271	0.1292	0.1314	0.1335	0.1357
−1.0	0.1379	0.1401	0.1423	0.1446	0.1469	0.1492	0.1515	0.1539	0.1562	0.1587
−0.9	0.1611	0.1635	0.1660	0.1685	0.1711	0.1736	0.1762	0.1788	0.1814	0.1841

$\frac{X_i - \mu}{\sigma}$	0.00	0.01	0.02	0.03	0.04	0.05	0.06	0.07	0.08	0.09
−0.8	0.1867	0.1894	0.1922	0.1949	0.1977	0.2005	0.2033	0.2061	0.2090	0.2119
−0.7	0.2148	0.2177	0.2207	0.2236	0.2266	0.2297	0.2327	0.2358	0.2389	0.2420
−0.6	0.2451	0.2483	0.2514	0.2546	0.2578	0.2611	0.2643	0.2676	0.2709	0.2743
−0.5	0.2776	0.2810	0.2843	0.2877	0.2912	0.2946	0.2981	0.3015	0.3050	0.3085
−0.4	0.3121	0.3156	0.3192	0.3228	0.3264	0.3300	0.3336	0.3372	0.3409	0.3446
−0.3	0.3483	0.3520	0.3557	0.3594	0.3632	0.3669	0.3707	0.3745	0.3783	0.3821
−0.2	0.3859	0.3897	0.3936	0.3974	0.4013	0.4052	0.4090	0.4129	0.4168	0.4207
−0.1	0.4247	0.4286	0.4325	0.4364	0.4404	0.4443	0.4483	0.4522	0.4562	0.4602
−0.0	0.4641	0.4681	0.4721	0.4761	0.4801	0.4840	0.4880	0.4920	0.4960	0.5000
+0.0	0.5000	0.5040	0.5080	0.5120	0.5160	0.5199	0.5239	0.5279	0.5319	0.5359
+0.1	0.5398	0.5438	0.5478	0.5517	0.5557	0.5596	0.5636	0.5675	0.5714	0.5753
+0.2	0.5793	0.5832	0.5871	0.5910	0.5948	0.5987	0.6026	0.6064	0.6103	0.6141
+0.3	0.6179	0.6217	0.6255	0.6293	0.6331	0.6368	0.6406	0.6443	0.6480	0.6517
+0.4	0.6554	0.6591	0.6628	0.6664	0.6700	0.6736	0.6772	0.6808	0.6844	0.6879
+0.5	0.6915	0.6950	0.6985	0.7019	0.7054	0.7088	0.7123	0.7157	0.7190	0.7224
+0.6	0.7257	0.7291	0.7324	0.7357	0.7389	0.7422	0.7454	0.7486	0.7517	0.7549
+0.7	0.7580	0.7611	0.7642	0.7673	0.7704	0.7734	0.7764	0.7794	0.7823	0.7852
+0.8	0.7881	0.7910	0.7939	0.7967	0.7995	0.8023	0.8051	0.8079	0.8106	0.8133
+0.9	0.8159	0.8186	0.8212	0.8238	0.8264	0.8289	0.8315	0.8340	0.8365	0.8389
+1.0	0.8413	0.8438	0.8461	0.8485	0.8508	0.8531	0.8554	0.8577	0.8599	0.8621
+1.1	0.8643	0.8665	0.8686	0.8708	0.8729	0.8749	0.8770	0.8790	0.8810	0.8830
+1.2	0.8849	0.8869	0.8888	0.8907	0.8925	0.8944	0.8962	0.8980	0.8997	0.9015
+1.3	0.9032	0.9049	0.9066	0.9082	0.9099	0.9115	0.9131	0.9147	0.9162	0.9177
+1.4	0.9192	0.9207	0.9222	0.9236	0.9251	0.9265	0.9279	0.9292	0.9306	0.9319
+1.5	0.9332	0.9345	0.9357	0.9370	0.9382	0.9394	0.9406	0.9418	0.9429	0.9441
+1.6	0.9452	0.9463	0.9474	0.9484	0.9495	0.9505	0.9515	0.9525	0.9535	0.9545
+1.7	0.9554	0.9564	0.9573	0.9582	0.9591	0.9599	0.9608	0.9616	0.9625	0.9633
+1.8	0.9641	0.9649	0.9656	0.9664	0.9671	0.9678	0.9686	0..9693	0.9699	0.9706
+1.9	0.9713	0.9719	0.9726	0.9732	0.9738	0.9744	0.9750	0.9756	0.9761	0.9767
+2.0	0.9773	0.9778	0.9783	0.9788	0.9793	0.9798	0.9803	0.9808	0.9812	0.9817
+2.1	0.9821	0.9826	0.9830	0.9834	0.9838	0.9842	0.9846	0.9850	0.9854	0.9857
+2.2	0.9861	0.9864	0.9868	0.9871	0.9875	0.9878	0.9881	0.9884	0.9887	0.9890
+2.3	0.9893	0.9896	0.9898	0.9901	0.9904	0.9906	0.9909	0.9911	0.9913	0.9916
+2.4	0.9918	0.9920	0.9922	0.9925	0.9927	0.9929	0.9931	0.9932	0.9934	0.9936
+2.5	0.9938	0.9940	0.9941	0.9943	0.9945	0.9946	0.9948	0.9949	0.9951	0.9952
+2.6	0.9953	0.9955	0.9956	0.9957	0.9959	0.9960	0.9961	0.9962	0.9963	0.9964
+2.7	0.9965	0.9966	0.9967	0.9968	0.9969	0.9970	0.9971	0.9972	0.9973	0.9974
+2.8	0.9974	0.9975	0.9976	0.9977	0.9977	0.9978	0.9979	0.9979	0.9980	0.9981
+2.9	0.9981	0.9982	0.9983	0.9983	0.9984	0.9984	0.9985	0.9985	0.9986	0.9986
+3.0	0.99865	0.99869	0.99874	0.99878	0.99882	0.99886	0.99889	0.99893	0.99896	0.99900
+3.1	0.99903	0.99906	0.99910	0.99913	0.99915	0.99918	0.99921	0.99924	0.99926	0.99929
+3.2	0.99931	0.99934	0.99936	0.99938	0.99940	0.99942	0.99944	0.99946	0.99948	0.99950
+3.3	0.99952	0.99953	0.99955	0.99957	0.99958	0.99960	0.99961	0.99962	0.99964	0.99965
+3.4	0.99966	0.99967	0.99969	0.99970	0.99971	0.99972	0.99973	0.99974	0.99975	0.99976
+3.5	0.99977	0.99978	0.99978	0.99979	0.99980	0.99981	0.99981	0.99982	0.99983	0.99983

Appendix B Sample examination questions

1. What are the main dimensions of quality management? Give examples of improvement methods which can be used for each of the dimensions.

   What are the main business benefits of total quality management (TQM)? How does the cost of quality for a manufacturing company change during each of the stages of implementing TQM. Why do so many companies experience difficulties in implementing TQM and how can these difficulties be overcome?
2. Describe the scope of each of the three parts of the ISO 9000 Standard. What are the main contributions made by the implementation of ISO 9000 towards the quality development of a manufacturing company?

   What are the main requirements of ISO 9001 for each of the following:

   (i) purchasing;
   (ii) storage;
   (iii) inspection.

   In which areas of quality system development is the use of statistical techniques appropriate? Describe how you would introduce acceptance sampling as part of an ISO 9000 implementation programme.
3. What is the role of failure mode effects analysis (FMEA) in the implementation of statistical process control? What are the stages in the creation of a control plan and describe the criteria used in the identification of critical characteristics?

   Components are manufactured on an automatic process for which the design specification for the length of the component is 30.5±0.5 mm.

   The data tabulated below represents the results of three different process trials. Evaluate the capability of the process in each case and decide which of the trials represents the most capable setting of the process.

Sample No.	Trial 1		Trial 2		Trial 3	
	Average length (mm)	Sample range (mm)	Average length (mm)	Sample range (mm)	Average length (mm)	Sample range (mm)
1	30.2	0.6	31.6	0.4	32.0	1.2
2	31.5	0.8	30.1	1.2	29.8	0.8
3	29.8	1.2	27.6	0.1	31.2	1.4
4	29.4	0.9	32.1	0.8	30.6	1.6
5	30.8	1.8	28.2	0.2	29.4	0.6
6	31.2	0.4	30.8	1.0	30.8	1.8

Each sample has five readings and for $n = 5$ the value of the constant d_n is 2.33.

4. What are the main stages in the application of quality function deployment (QFD)? What are the benefits from using QFD in the development of quality products? What is the role of design of experiments (DOE) in the development of quality products?

   A manufacturing process has two critical parameters, A and B, which can both be set at either high or low settings. The table below represents the L_4 orthogonal array and indicates the yield of the process (in percentage conversion) for each of the settings of both A and B.

Experiment	Parameter A	Parameter B	Yield (%)
1	High	High	68
2	High	Low	54
3	Low	High	52
4	Low	Low	70

   What are the optimum settings of the process parameters A and B? how would this experimentation be extended to evaluate the effects of a third parameter, C, which can also be set at either a high or low level? How would you test whether an experimental array is orthogonal?

5. Describe how characteristic reliability functions can be used to determine the reliability of components combined together to form systems. How can the reliability of systems be improved? What effect does the periodic maintenance of components have upon the reliability of systems and the overall cost of operations?

   A manufacturing process is configured as show in Figure B1. Products are manufactured by initially undergoing process A before being transferred via a transfer line to either of the process B stations before finally being processed at process C.

   Assuming that each of the elements of the system can be modelled using the exponential model determine the reliability of the system at 500 hours.

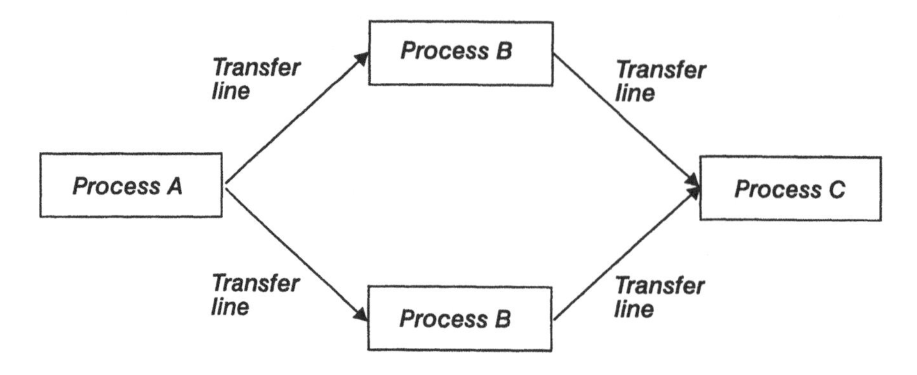

Figure B1

Element	Age specific failure rate (10^{-4} hours)
Process A	3.25
Process B	8.62
Process C	1.45
Transfer Line	2.79

6. Describe what is meant by the term 'quality system' and discuss the role of the International Standard ISO 9000 series in the development of a quality system within a manufacturing company.

   Describe in detail the requirements of ISO 9000 in each of the following areas:

   (i) management responsibility;
   (ii) quality system;
   (iii) internal audits.

   What are the main stages in the auditing of a quality system and what are the main activities undertaken at each stage? What is the role of internal auditing in quality system development?

   For a company which has implemented ISO 9000 what are the main activities involved in moving towards being a total quality management (TQM) company? Discuss the benefits of implementing ISO 9000 before embarking upon TQM.

7. (a) Where is acceptance sampling most appropriate in a manufacturing company? Describe two different ways in which a sampling plan can be established and give details of the data required in each case.

   Discuss the influence of each of the following on the design of sampling plans:

(i) production method;
(ii) value added by a process;
(iii) method of testing.

(b) Describe the role of statistical process control (SPC) in the development of quality management within a manufacturing company. What are the main indicators provided by SPC of the quality performance of a company?

Describe the role of each of the following in the implementation of SPC:

(i) failure mode effects analysis;
(ii) process capability studies;
(iii) selection of the type of control chart.

What are the confidence levels associated with the control limits on means and range charts for variables?

8. Discuss the Taguchi concept of the quality loss function and describe the potential use of this function in the management of quality for a manufacturing company. How does the Taguchi approach to quality costing compare with the prevention/appraisal failure cost model adopted by BS 6143?

A food manufacturer is producing a soft drink which is required to have a sugar content of 1.1 g ± 0.2 g which is based upon consumer preference for the correct nominal sweetness. The cost of replacing a drink to the consumer which is out of specification is £1.80. The company have the facility for diluting or concentrating the sugar content but this additional process will add £0.20 per drink.

Determine the appropriate manufacturing tolerance for the sugar content of the drink.

9. Describe the role of the characteristic reliability functions in the implementation of preventative maintenance.

How would you establish the maintenance periodicity for an item of process equipment?

What are the considerations you would take into account when deciding upon an appropriate reliability model?

A company's maintenance programme requires that a critical electric motor is periodically maintained. The key components in the reliability of the motor are two identical support bearings, one situated at the drive end of the motor and the other at the non-drive end. If either bearing fails, the motor is rendered inoperable.

Using Weibull paper, determine from the data given below the maintenance period for replacing the bearings to ensure the reliability of the motor is always greater than 50%. Comment upon your result.

% of individual bearings failing	Time to failure (hours)
20%	2000
30%	4000
40%	7000

10. Describe the difference in scope between each of the three parts of ISO 9000. Which part of the standard is most applicable in each of the following and why.

(i) a small shipyard?
(ii) a high volume consumer electronics manufacturer?
(iii) a pressure vessel repair company?

Outline the requirements of ISO 9002 for each of the following:

(i) Document control;
(ii) process control;
(iii) training.

What are the main stages in the implementation of ISO 9000 for a manufacturing company? What are the main resource implications of such a programme?

Discuss the relationship between ISO 9000 implementation and the implementation of total quality management. What affect would each of these activities have on the cost of quality for a manufacturing company?

11. What are the main stages involved in the implementation of the following statistical quality control techniques:

(i) statistical process control?
(ii) acceptance sampling by attributes?
(iii) Taguchi methods?

Describe how you would use these three techniques as part of a quality improvement programme.

What factors influence the selection of the most appropriate form of process control chart?

Where are sampling plans most appropriately used and what are the criteria that should be used in the selection of sampling plans?

How would you select the manufacturing process data to be used in the Taguchi matrix? What factors would you take into account in selecting the data?

12. What are the characteristic reliability functions and how can these be used in maintenance management? How would you determine these characteristic reliability functions for a piece of equipment undergoing periodic maintenance?

A process plant is arranged as shown in Figure B2 with each of the processes A to C undergoing planned maintenance. Maintenance data has shown the individual processes to have the following age-specific failure rates:

Process	**Age specific failure rate (10^{-3} hr)**
A	5.73
B	4.33
C	3.19

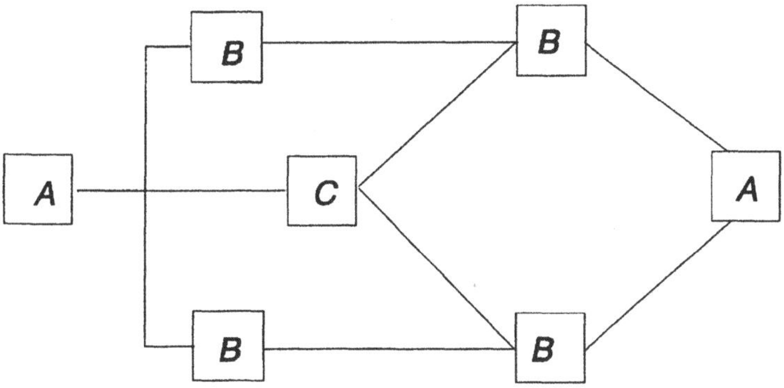

Figure B2

Determine the reliability of the complete manufacturing system at 160 hours assuming each process can be modelled using the exponential model. Given that the planned period between process maintenance is 160 hours comment upon your result.

13. What is the role of the quality management function within a manufacturing company and how does this change with the implementation of total quality management (TQM)?

Describe the use of the following techniques in the implementation of TQM:

(i) quality costs;
(ii) supplier assessment;
(iii) process control charts;
(iv) quality improvement teams.

In each case explain how you would implement the technique as part of a TQM programme.

14. Describe the use of advanced quality planning techniques in modern manufacturing management.

What are the main benefits associated with the following:

(i) failure mode effects analysis
(ii) Taguchi methods

in the quality improvements process? How would you implement each of these techniques?

A manufacturing company has established a process capability requirement of 1 for new process equipment. A new machining centre is commissioned and a pilot production run produces the following 12 samples (sample size = 4).

Sample no.	**Sample mean (mm)**	**Sample range (mm)**
1	103	8
2	101	2
3	96	6
4	101	5
5	100	11
6	98	2
7	95	3
8	103	8
9	103	6
10	96	6
11	99	7
12	100	2

(For sample size 4, $d_n = 2.059$.)

Determine the value of process capability, C_P and process capability index C_{Pk} given that the design specification is 101 ± 8 mm. Comment upon your result in terms of process improvement and draw a cusum plot of the pilot run results.

15. Describe how you would determine the reliability of a manufactured product. What are the main stages involved in the collection and analysis of the data necessary to determine product reliability? What factors would influence the accuracy of the results produced?

Forty manufactured products are produced over a period of time and the following failure date is collected:

Cumulative number failing	**Age at failure (hours)**
4	18.5
8	27
16	40
32	70

Using Weibull paper, determine the reliability of the product at 80 hours. Given that this product has a warranty period of 80 hours, comment upon your result and identify any corrective actions you would recommend.

16. What are the main features of acceptance sampling. Describe how you would implement sampling in the following manufacturing environments:

- large volume production of low cost components requiring a defined maximum defect rate;
- precision components for the aerospace industry having high inspection costs and produced on reliable production processes.

Batches of components are to be submitted for acceptance sampling. The sampling plan is required to provide a probability of acceptance of 0.999 for batches at the acceptable quality level of 1% and a probability of 0.01 that batches at the lot tolerance per cent defective of 8% will be accepted.

Design a single sampling plan using a modified Thorndike chart to meet these requirements. What guarantees can be given regarding the maximum defect level in outgoing batches.

17. Describe how failure mode effects analysis can be used to determine process reliability and the key characteristics for the application of statistical process control.

What criteria would you use in the selection of the most appropriate type of process control chart? How would you determine 'process capability' and how would you use this measure to establish product reliability?

A manufacturing system comprises two sequential production processes, one of which has a failure rate of 5×10^{-3} and the other a failure rate of 2.7×10^{-3}. The system can be represented using the exponential model and is the subject of a maintenance programme operating at 160 hour intervals. What is the reliability of the system at the end of the maintenance period and how does this compare with the reliability at the mean time to failure? Comment upon your result.

18. What are the stages involved in the implementation of statistical process control in a manufacturing company?

Describe how you would select the most appropriate type of process data for establishing process control charts.

Which type of control chart would you employ in each of the following manufacturing situations:

(i) a process producing a continuous length of product;
(ii) the production of low cost components made in large batches;
(iii) the small batch production of precision components.

Give explanations of your choice in each case.

During a pilot study, with the process in a known state of control, components are produced and details of the 10 samples taken (each with a sample size of 4) are shown below. The design tolerance for this component is 100 mm ± 10 mm.

Sample no.	Sample mean value (mm)	Sample range (mm)
1	107	28
2	101	10
3	98	37
4	99	20
5	109	26
6	102	25
7	97	18
8	105	31
9	108	22
10	104	25

Produce appropriate mean and range charts based upon these results and comment upon the process capability.

Constants

d_n = 2.059 (for sample size, $n = 4$)
D_1 = 0.19
D_2 = 0.29
D_3 = 1.93
D_4 = 2.57

19. Discuss the significance of the reliability of a manufactured product in relation to the customer's perception of 'quality'.

Describe how you would determine the reliability of a new product and how would you expect its performance to vary with time.

A manufactured product is assembled from components as shown in Figure B3. Assuming these components can be modelled using the exponential model and that the suppliers age test some components as indicated below, what is the reliability of the system at 200 hours given the following information:

Component	Age supplied	Age specific failure rate (10^{-4} hours)
A	New	11.16
B	50 hours	45.81
C	Unknown	11.16

Given that the end customer is to be provided with a 200 hour warranty on this product, comment upon your result and any actions you would prescribe.

20. What are the cost factors that should be considered when deciding whether to adopt statistical sampling as an approach to product

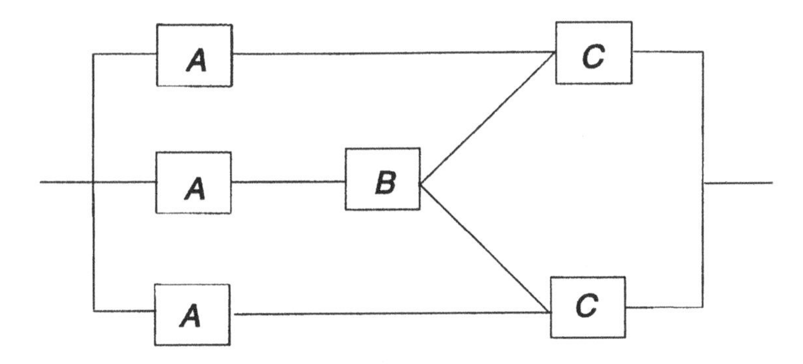

Figure B3

inspection? Give examples which clearly indicate where sampling would be appropriate and also where it would be inappropriate.

Describe how you would implement a sampling plan in a manufacturing company with particular reference to (i) selecting sample sizes; (ii) establishing acceptable quality levels; and (iii) determining sampling frequency.

A company wishes to adopt statistical sampling to BS 6001 using an acceptable quality level (AQL) of 4% with normal inspection level II. The product manufactured by the company can be made in batch sizes of either 10 or 100. Using the data from BS 6001 given below calculate for each batch size the average outgoing quality when production is running at the AQL. For each case determine the average outgoing quality limit. As quality manager for this company which batch size would you select, giving reasons for your choice?

Data from BS 6001

For single sampling to normal inspection level II, Tables I and IIA give:

Batch size	**Sample size code**	**Sample size**
9 to 15	B	3
501 to 1200	J	80

For an AQL = 4%, Tables x – B and x – J give:

Percentage of lots expected to be accepted (Pa)	**Percent defective**	
	Sample code B	**Sample code J**
99	0.33	3.72
95	1.70	5.06
90	3.45	5.91
75	9.14	7.50
50	20.60	9.55

Percentage of lots expected to be accepted (Pa)	Percent defective Sample code B	Sample code J
25	37.00	11.90
10	53.60	14.20
5	63.20	15.80
1	78.40	18.90

21. Describe how you would use the characteristic reliability functions in the management of maintenance costs.

Explain how you would use such functions to model the reliability of process equipment which is periodically maintained.

A machine shop has 50 machining centres with each machine undertaking all the machining operations necessary to produce a product. Using the data given below, determine the reliability of machining centres at 450 hours using Weibull paper. Comment upon your results.

As an alternative to undertaking all the machining operations for each part on one machine, the production planning department are considering grouping the machines into cells. Each cell would comprise three machines processing the parts in sequence. The effect of this change would be to reduce the machine set-up times and thereby increase the overall availability of the three machines from 60% to 85%.

Determine whether the reliability at 450 hours of the proposed cellular configuration makes this regrouping a viable proposition. Given reasons for your decision.

Data:

Cumulative number of machining centres having failed	Time to failure (hours)
5	200
9	400
20	1000
25	1400

22. What are the main features of sampling plans and where would you most appropriately apply statistical sampling to the manufacturing process?

A company operating a complex production process requires a single sampling plan which will achieve the 2% acceptable quality level specified by its major customer.

The practical problems associated with in-process sampling means that the maximum sample size that can be taken from the batches of 250 is 30. Using a modified Thorndike chart design a sampling plan that will give a probability of 0.95 of accepting a batch at the acceptable quality level.

Based on this sampling plan, what guarantees can be given to the customer regarding the outgoing quality level?

23. (a) Discuss the scope of the ISO Standard 9000 used for the third-party evaluation of a company's quality assurance system. Describe the stages involved in the implementation within a small manufacturing company of a quality assurance programme to comply with this Standard.

(b) What are the two forms of control charts constructed from attribute data? Indicate in your answer how such charts are constructed and give an example of what would be an appropriate industrial application of each.

24. What are the costs associated with the process quality level and reliability for a manufacturing company?

A large machine shop has installed 20 new machining centres and has collected the following maintenance data relating to the operating time to failure of these machines:

Time to failure (hours)	Cumulative number failed
200	1
850	4
2000	8
6250	16

Based on this information, use Weibull paper to determine the reliability of the machining centres at 1000 hours.

Some new machining centres (M) with the same reliability are to be arranged into a production cell together with a robot handling device (R). This production facility can be represented in terms of a Bayes circuit as shown in Figure B4.

Given that the robot has a reliability of 0.9 at 1000 hours, use Bayes' theorem to determine the reliability of the system at 1000 hours.

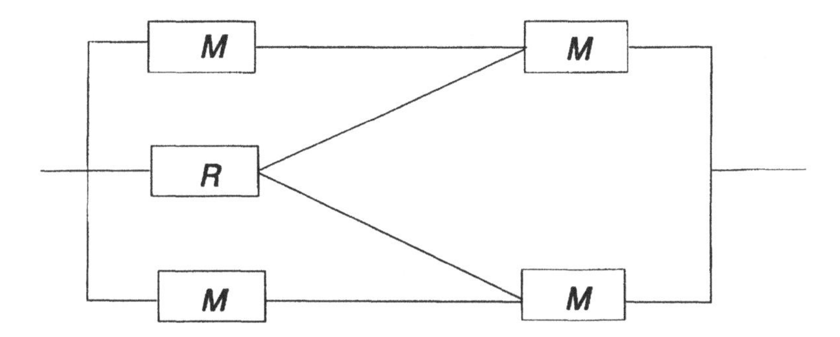

Figure B4

Appendix C Outline solutions to sample examination questions

1. The main dimensions of Quality management covered are:

- people;
- systems;
- tools and techniques.

Examples of improvement methods include:

- teamworking, empowerment etc.;
- ISO 9000 and Q101 etc.;
- SPC, problem solving tools, QFD etc.

The main business benefits of TQM are:

- reduced cost of quality;
- improved customer service and retention;
- reduced process and development lead times.

The changes in the cost of quality include:

Stage 1	**Stage 2**	**Stage 3**
Systems orientation	Improvement	Prevention
20% ——>	15% ——>	8% ——>5%

Typical distribution of costs, investment in prevention rather than failure.

Problems in implementing TQM:

- improving all three dimensions simultaneously;
- regenerating ongoing continuous improvements;
- sustaining management commitment.

Overcome by using:

- combination of systems, techniques and people development;

- changing the focus of TQ implementation;
- long-term senior management commitment.

2. Scope of the main elements of ISO 9000 are:

- 9001 – Design, Manufacture and Test;
- 9002 – Manufacture and Test;
- 9003 – Test.

Main contributions are:

- provide basic systems foundation and definition of requirements;
- gives companywide view of quality;
- externally 'policed' quality commitment.

Requirements of ISO 9000 are:

- purchasing: requires formalized procedure, assessment of subcontractors, definition of requirements, verification of materials;
- storage: formalized procedures, control of issue and use, protection and preservation, designated areas, labelling and identification;
- inspection: formalized procedures, verification of incoming materials, in-process inspection, final approval/positive release, records of inspection and acceptance criteria.

Quality systems may include statistical techniques in the following areas:

- acceptance sampling to control incoming materials (4.10.2) in-process inspection (4.10.3), final inspection (4.10.4) and process control (4.9);
- SPC in process control (4.9).

Introduction of acceptance sampling should include:

- definition of procedures and criteria;
- adoption of BS 6000.

3. FMEA is used in implementing SPC for:

- identification of critical characteristics (RPN);
- measurement of capability;
- measurement of improvement through RPN.

Stages in creating a control plan:

- identify critical characteristics from FMEA;
- identify acceptance criteria and sampling;
- define process capabilities and control limits.

Criteria for identifying critical characteristics:

- severity, occurrence and detection;
- limiting criteria for RPN.

Calculation

$$\text{Process capability, } C_p = \frac{\text{Tolerance}}{6\sigma_{\bar{X}}} \text{ where } \sigma_{\bar{X}} = \frac{\overline{W_X}}{d_n\sqrt{n}}$$

	Trial 1	Trial 2	Trial 3
$\bar{X}$	30.48	30.07	30.63
$\overline{W_X}$	0.95	0.62	1.23
$6\sigma_{\bar{X}}$	1.09	0.71	1.42
C_p	0.91	1.41	0.70

Trial 2 has best capability but poor C_{pk} hence would need resetting.

4. Main stages in applying QFD are:

- four main matrices;
- relationship between customer requirements and technical features;
- rating and ranking of features.

Benefits of QFD:

- team-based inputs to product development;
- clear linkage between requirements and functionality;
- capturing the 'voice of the customer'.

Role of DOE should include:

- improved robustness of product design;
- improved signal to noise ratio;
- optimization of process settings.

Calculation

Parameter A:	high setting ave. yield	=	61%
	low setting ave. yield	=	61%
Parameter B:	high setting ave. yield	=	60%
	low setting ave. yield	=	62%

Hence optimum settings are B_{Low} and $A_{High\ or\ Low}$ (hence another criterion, e.g. cost, would be used to select parameter A setting).

Matrix for C is

A	B	C	
H	H	H	
H	L	H	Orthogonal if
H	H	L	ratio of B_1:B_2 for A_1
H	L	L	is equal to B_1:B_2 for A_2 etc.
L	H	H	
L	L	H	
L	H	L	
L	L	L	

5. Reliability functions used to determine systems reliability are as follows:

- models used to establish component reliability (e.g. exponential or Weibull);
- component reliabilities used to analyse systems as series, parallel, standby or Bayes;
- systems reduced to a simple series or parallel arrangement.

Systems reliability improvement:

- introducing parallel systems (where functionality is not altered);
- standby systems;
- improving individual component reliabilities.

Periodic maintenance provides:

- maintenance prior to MTBF reduces probability of failure for components and hence improves system reliability;
- maintenance costs (prevention) incurred to reduce breakdown costs (failure).

Calculation

Component reliabilities: $R_A = 0.85$ $R_C = 0.93$
at 500 hours using $e^{-\lambda t}$ $R_B = 0.65$ $R_{T\text{-}L} = 0.87$

system reduces to:

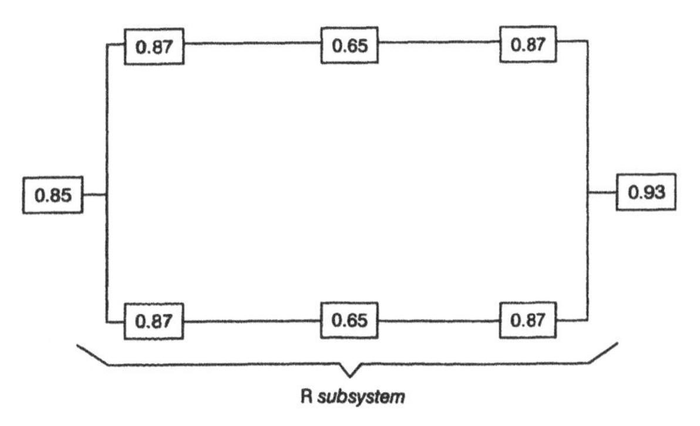

$R_{subsystem} = 0.87 \times 0.65 \times 0.87 = 0.49$
$R_{system} = R_A \times [1 - (1 - R_{subsystem})^2] \times R_C = 0.57$ (poor)

6. Description of the term 'quality system' should include:

- all activities from purchasing through manufacturing to inspection and despatch;
- definition of responsibilities, processes and records;
- concepts of system design, implementation and maintenance.

The role of ISO 9000 series:

- scope of the standard in terms of 9001, 9002 and 9003;
- provides a framework in terms of sections 4.1. to 4.20 for the requirements of a quality system;
- provides for third-party assessment and auditing which is internationally accepted.

The detailed description of sections 4.1, 4.2 and 4.16 should include:

(i) definition of responsibility (for items 4.1.2.1 (a)–(e) in ISO 9000), verification resources, management representative and management review;
(ii) requirements for system documentation, quality plans, quality manual and procedures;
(iii) independence of auditors, training of auditors and audit record, review and close-out.

The main stages/techniques of auditing covered are:

(a) audit planning and preparation – involving document review, audit checklists, auditee notification;
(b) audit execution – involving entry/exit arrangements identification of NCs, CARs and audit conduct;
(c) audit follow-up – close-out of CARs, audit reporting and management review.

Role of internal auditing on development:

- mechanism for ensuring effective corrective action (4.13);
- forms part of the management review;
- independent 'policing' of the internal systems.

Stages in development covered are:

Continuous Improvement

Systems orientation	⇒	Improvement orientation	⇒	Prevention orientation
• *ISO 9000*		• *Team based Problem solving*		• *Advanced quality planning*

Benefits of adopting ISO 9000 initially are:

- it provides a reliable systems 'foundation';
- procedures provide a mechanism for capturing improvements;
- first stage in workforce orientation and training.

7. (a) Acceptance sampling is appropriate where:

- a change of responsibility;
- destructive testing;
- high/defect processes (after);
- high added value (before);
- in-line processes (before);

The two approaches to sampling plan design covered are:

- Thorndike charts based upon batch size, AQL and LTPD;
- BS 6001 based upon lot size, sampling level and AQL.

The discussion on the effect on the design of sampling plans should include:

(i) batch production, single or multiple plans depending upon process variance, process production, sequential plans;
(ii) cost of inspection vs cost of failure using producer's and consumer's risk hence affecting sample size and acceptance number;
(iii) cost of inspection, disruption to process vs risk of batch rejection affecting the selection of multiple or single or sequential.

(b) Description of the role of SPC should include:

- improvement in process control through reduced variability;
- 6 σ requirement for suppliers;
- operator involvement in the improvement process.

The main indicators provided by SPC are:

- process capability C_p;
- process capability index $C_{Pk;}$
- process effective range $\pm 3\sigma$;
- trend analysis and cyclic phenomenon.

The roles of the techniques are as follows:

(i) FMEA defines the critical characteristics through the calculation of failure.
(ii) Process capability studies for CP < 1 require improvement, CP >> 1 require no action and CP ~ 1 require SPC.
(iii) Control chart type includes an evaluation of variables vs attribute data, lot size and sample size.

The confidence levels are:

- upper and lower warning limits at $\pm 2\sigma$ are 95%
- upper and lower action limits at $\pm 3\sigma$ are 99.7%

from the normal distribution.

8. Discussion on the loss function should include:

- the concept of product quality being the (minimum) loss imparted to society;
- changes in traditional thinking which relates costs purely to achieving tolerance;
- achieving uniformity actually reduces the cost of quality, a basic tenant of total quality management.

Comparison with PAF costing should include:

- BS 6143 relates to the overall cost of quality for a company, rather than product costing;
- Taguchi uses the concept of loss to society, rather than simple external failure costs;
- BS 6143 addresses systems costing whereas Taguchi is product based;
- Taguchi can, however, be seen to be promoting the prevention-orientated approach to quality.

Main types of quality characteristics covered are:

- nominal-the-best
- smaller-the-better
- larger-the-better

} include examples.

Calculation

First part requires the calculation of the Taguchi proportionality constant from:

$$L(y) = K(y-m)^2$$

$$K = \frac{A_0}{\Delta_0^2} = \frac{1.8}{0.2^2} - £45/\mathrm{g}$$

Hence the value of the characteristic y when the loss is £0.20 is:

$$y = \sqrt{\frac{A}{K}} = \sqrt{\frac{0.2}{45}} = 0.067\ \mathrm{g}$$

Hence the manufacturing specification would be:

$$1.1 \text{ g} \pm 0.067 \text{ g}$$

9. Characteristic reliability functions covered in the text include:

- probability density function (pdf) ($f(t)$);
- reliability characteristic function ($R(t)$);
- unreliability function ($F(t)$);
- failure rate $\lambda(t)$;
- MTTF ($E(t)$).

Used to establish plant reliability and hence maintenance requirements.

Maintenance periodicity can be established as follows.

- Determine reliability characteristic using appropriate model (e.g. exponential, Weibull).
- Establish availability requirement over specified timescale.
- Plan maintenance period and monitor ongoing performance for model change.

Factors to take into account:

- failure period – early life, random or wear-out;
- accuracy requirement and data availability;
- failure costs vs maintenance cost;
- systems models for series, parallel or standby operation.

Calculation from Weibull plot:

- shaping constant $\beta = 0.6$ (this employs early life failure);
- characterisitc life $\eta = 21\,000$ hours.

From the Chartwell expression:

$$R(t) = \exp(-[\frac{t - t_0}{\eta - t_0}]^{\beta})$$

for $R(t) = 0.5$, assuming $t_0 = 0$. Hence $t = 11\,500$ hours (approx.). However, for 2 bearings, to provide $R(t) = 0.5$, hence *each* bearing to have $R(t) = 0.707$. For $R(t) = 0.707$, assuming $t_0 = 0$ Hence $t = 3600$ hours (approx.).

10. Description of Part 1 (design, manufacturing and test), Part 2 (manufacturing and test) and Part 3 (test):

(i) Part 1: ISO 9001;
(ii) Part 2: ISO 9002;
(iii) Part 2: ISO 9002.

Detailed requirements for:

(i) document control contains 4.5.1 Document approval and issue, 4.5.2 Document change/modification;
(ii) process control contains 4.9.1 Work instructions and workmanship criteria, 4.9.2 Special processes;
(iii) Training 4.18 requires training needs analysis, appropriate qualification and training records.

Main stages of ISO 9000 implementation:

- identification of existing systems;
- preparation of documented policy and procedures;
- implementation of controls and records (material control, process records, calibration, etc.);
- audit, review and accreditation.

Main resources:

- management time and commitment;
- stores, calibration, process controls, etc.

Relationship between QA and TQM is shown in Figure C1.

Costs of quality:

(i) ISO 9000 incurrs appraisal costs and prevention costs (specifications and vendor control)
(ii) TQM involves investment in prevention through improvement activities.

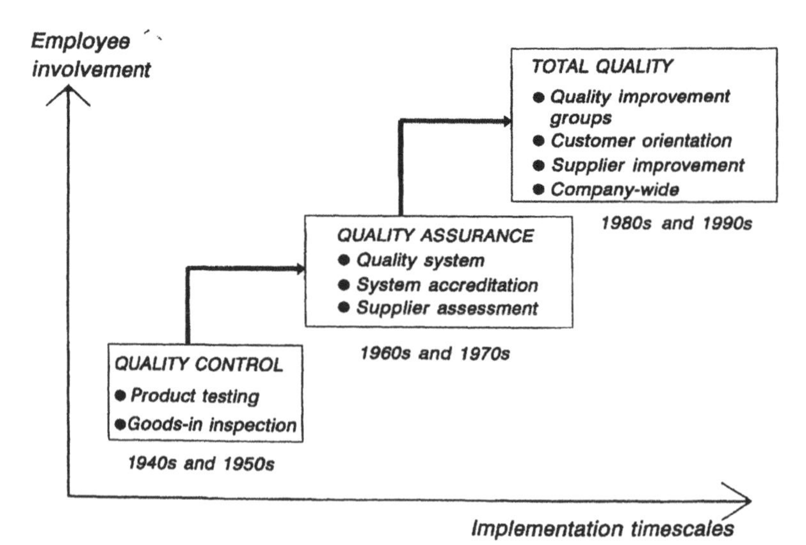

Figure C1 Relationship between QA and TQM.

11. Stages involved in:

(i) SPC:
- define critical variables (FMEA);
- establish process capability;
- pilot study for σ;
- implement charts at $\pm$ 2 and 3 σ;
- define procedures for out of control.

(ii) Acceptable sampling using:
- Thorndike charts based on α, β, AQL and LTPD;
- BS 6001.

(iii) Taguchi:
- multi-vari experimental matrices;
- identification of key variables;
- control limits for key variables.

For quality improvement:

- techniques used by improvement teams;
- part of an overall improvement methodology for diagnosis (SPC and Taguchi) and remedial (sampling) cycles;
- used as improvement measurement techniques.

Factors for control chart selection:

- attributes vs variable data;
- cost of analysis vs cost of sampling;
- means or range or trend analysis required.

Appropriateness of sampling plans:

- input and output items;
- where 100% inspection not appropriate;
- cost implications.

Criteria are:

- producer's vs consumer's risk;
- acceptable and unacceptable quality levels.

Taguchi matrix parameter selection:

- critical, measurable variables;
- first or second order measures;
- customer critical parameters (QFD).

12. Characteristic functions include:

- probability density function $f(t)$;
- reliability function $F(t) = \int f(t)$;

- unreliability function $R(t) = 1 - F(t)$.

For maintenance to determine:

- MTTF and MTBF;
- individual machine failure prediction.

To determine reliabilities:

- model using either exponential or Weibull models;
- collect failure data and plot failure vs cycles;
- calculate failure constants (λ or η).

Calculation

Reliabilities of processes A to C using $R = e^{-\lambda t}$: $R_A = 0.4$
$R_B = 0.5$
$R_C = 0.6$

Process elements from Figure B1 are:

R Bayes = R_C [System C is good] + [$1 - R_C$] [System C is bad]

Hence R system = $R_A \times$ R Bayes $\times R_A$.

Include comments on poor overall system reliability governed by process A (reduce maintenance period from 160 hours).

13. Discussion of traditional role of quality management as integrator (as shown in Figure C2)

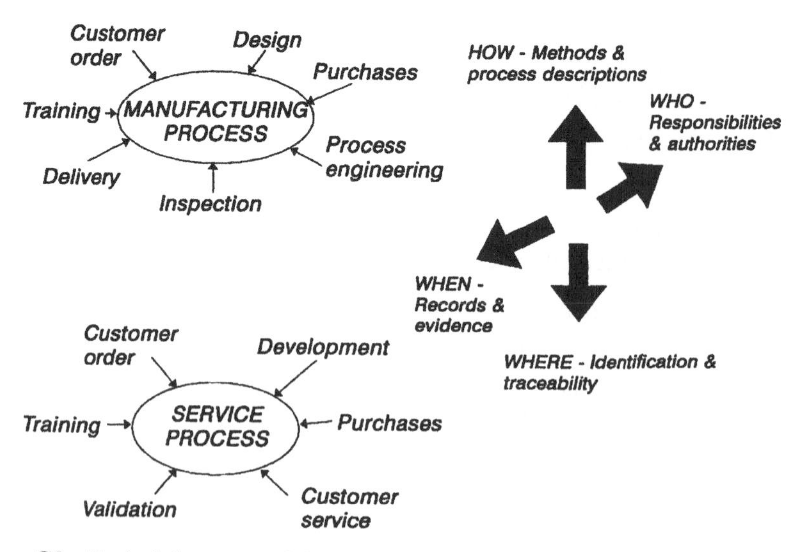

Figure C2 Typical elements and dimensions of a quality system.

between departments primarily as a control or assurance-based function. Changes involved in TQM should include:

- management-led, company-wide interpretation of quality;
- customer focus and conformance to specification;
- quality improvement teams and change in management style.

Implementation of the techniques should include the following.

(i) Quality costs collected in terms of prevention, appraisal and internal/external failure per as BS 6143 used as a basis for improvement measurement – focus on prevention. Categories of costs should be identified.

(ii) Supplier assessment including third party accreditation using ISO 9000/BS 5750 and the scope of such assessment, i.e. contract revision → design control → purchasing → process control → inspection → corrective action → storage/handling, etc. Cost of ownership and focus on specification.

(iii) Process control implementation via identification through pilot study of C_p and C_{Pk}. Identify process is under control then employ control charts UCL/LCL $\pm$ 3σ. Operation of preventative approach to the control of process variability. Widespread application in TQM.

(iv) QITs operating company-wide (deployed teams and quality circles) to focus on improvement projects. Description of types of teams, organizational structure through quality improvement councils and facilitations.

14. Discussion of advanced quality planning techniques covering:

- FMEA;
- DOE;
- QFD;
- control plans.

Provide an emphasis on prevention to reduce product development lead times and total lifecycle cost of quality.

(i) FMEA implemented through the identification of process → failure models → causes → effects → controls → RPN → corrective action and the team-based approach to product introduction. Benefits include reduction in failure costs and prioritizing of critical characteristics for control.

(ii) Taguchi relates the design parameters to the noise parameters to determine manufacturing tolerance to variability. Design of experiments implemented through relating design variables to process variables and establishing design and noise matrices.

Calculation

$$\text{Mean of sample range} = \frac{66}{12} = 5.5$$

$$\text{Hence } \sigma = \frac{\overline{W}}{d_n} = \frac{5.5}{2.059} = 2.67$$

$$\text{Also } \overline{\overline{X}} = 99.58$$

$$\text{Hence } C_p = \frac{\text{Design tolerance}}{6\sigma} = \frac{16}{6 \times 2.67} \approx 1$$

$$\text{Also } C_{Pk} = \frac{\text{Nearest limit } - \overline{\overline{X}}}{3\sigma} = 0.82$$

15. Determination of reliability using characteristic probability function appropriately modelled using:

- exponential model, assuming $R(t) = e^{\lambda t}$ only appropriate where random failure is occuring (i.e. $\lambda(t)$ = constant);
- Weibull distribution where:

$$R(t) = \exp - [\frac{t}{\eta}]^{\beta}$$

determined using shaping constant β and characteristic life η.
Stages include:

- establish process in a state of control and monitor production run;
- ensure products are applied under controlled conditions with a defined start time (t_0);
- record failure times and number of failings;
- plot failure data on Weibull plot to determine β and η

Data accuracy determined by (a) control of conditions, (b) pre-defined start times, (c) manufacturing/defect control etc.

Calculation

The Weibull chart gives $\beta = 2$, $\eta = 55$ hours

$$\text{Hence } R(80) = \exp - [\frac{80}{55}]^2 = 0.12$$

The low reliability and the value of $\beta = 2$ indicates 'age' failure outside the design capability.

16. Main features of sampling plans:

- sample taken (sample size);

- acceptance number;
- random samples to be taken;
- failed batches 100% inspection/rejection;
- definable AOQ and AOQL.

Implementation:

(i) Attribute testing, single sampling based on AOQL defined at a given per cent defective. Key considerations are inspection costs vs failure costs.
(iii) Variables testing using sequential sampling against tolerances – potential for smaller sampling. Trend analysis (cusum) may also prove relevant.

Consider the sampling plans. From Thorndike chart:

c	$\frac{PD \times n}{100}$ at α	$\frac{PD \times n}{100}$ at α
1	:	:
2	:	:
3	:	:
4	:	:
5	:	:
6	:	:
7	$X1$	$X2$

Examination reveals ratio of $X1 : X2$ is 1 : 8, the ratio of AQL : LTPD. Hence $n = 200$ and $c = 7$.

The maximum defect for outgoing batches is defined by AOQL determined from the plot of AOQ vs PD.

17. FMEA is implemented as follows:

- identify process failure modes;
- identify failure causes and occurrence;
- estimate failure severity;
- assess detection probabilities;
- calculate risk priority number;
- identify key characteristics for SPC and develop control plan.

Criteria for the selection of most appropriate control chart:

- attribute vs variable data;
- means, range or trend charts;
- discussion of attribute (means/range) p- and c-charts;
- costs of inspection vs cost of sampling.

Process capability determined using:

- pilot study to determine process σ;

- then take design tolerance 6σ;
- product reliability established from capability index.

Exponential model $e^{-\lambda t}$ failure rate is constant hence reliability depends only upon the elapsed time.

R1 × R2	R1	= 0.65
	R2	= 0.45
	R system	= 0.29

Mean time to failure = e^{λ} = 0.37

Hence failure probability is higher than MTTF and therefore reliability (component 2 in particular) would need to be improved.

18. Stages in the implementation of SPC:

- select process characteristic to be controlled;
- conduct pilot study to determine x and σ;
- design control limits using the pilot data;
- check control limits to ensure they are realistic and economically feasible;
- take samples of process and plot characteristics on control chart;
- establish action and warning procedures.

Selection of process data is based upon:

- cost implications;
- process disturbance;
- criticality of parameters.

Choices include:

- attribute vs variables data,
- inspection costs vs sampling cost;
- range/means charts vs proportion defective vs number, defective vs cusum.

Selection of appropriate data should include:

(i) no identifiable batch, defects per unit length (say) – C-chart;
(ii) sampling using low cost attribute data, batch size variable – p-chart;
(iii) consider both process drift and range – possibly need to consider small variation in means, mean, range or cusum.

$$\sigma = \frac{\overline{W}x}{d_n\sqrt{n}} = \frac{24.2}{2.059 \times 2} = 5.877 \text{ mm}$$

Also $\overline{X} = 103$ mm

Hence for means UAL $= \overline{X} + 3 \times \sigma = 120.6$

LAL $= \overline{X} - 3 \times \sigma = 85.37$

Hence for means UAL $= D_4 \times \overline{W} = 62.19$

LAL $= D_1 \times \overline{W} = 4.60$

Also process capability $C_p = \frac{\text{Design tolerance}}{6 \times \sigma} = 0.57$

Comments:

- Process capability is poor < 1, i.e. less than 1.
- Control chart action limits would not provide control within tolerances
- Mean of ranges is wider than design tolerance.

19. Product reliability is the performance of a product under the required operating conditions for a prescribed period of time.

Impact of perceived 'quality':

- conformance to a specification which usually relates to operating life;
- costs of warranty and replacement;
- market competition for reliable product.

Requires:

- clearly established requirements specification;
- knowledge of operating conditions,
- reliability characteristics of the product.

Reliability can be determined by examination of failure data (cumulative with respect to time). This data can then be used to:

- produce a Weibull plot of cumulative number of failures vs operating cycles (time),
- determine a straight line plot;
- from the Weibull plot, determine the scaling constant α and the shaping constant β:

$$R(t) = \exp(-[\frac{t - t_0}{\alpha}]^{\beta})$$

Most products would exhibit the classical performance/time relationship illustrated by considering the age-specific failure rate.

The Weibull shaping constant β would indicate the appropriate part of the curve.

Reliability is independent of initial age when modelled using the exponential function hence:

$$R_A = e^{-\lambda t} = e^{-11.16\times10^{-4}\times200} = 0.8$$

$$R_B = e^{-\lambda t} = e^{-45.81\times10^{-4}\times200} = 0.4$$

$$R_C = R_A \qquad = 0.8$$

$$\text{Hence } R_S = R_{Good} \times 0.32 + R_{Bad} \times 0.68 = 0.9$$

20. Cost factors include:

- cost of inspection vs cost of risk;
- risks include producers' and consumers' risks,
- inspection costs may be high due to damage, interference, delays involved, etc.,
- 100% inspection not appropriate for destructive testing;
- extreme alternative is zero testing – scrap costs and customer warranty costs problems;
- accuracy may be poor due to fatigue/boredom.

Sampling appropriate:

- before high added value, large batch processes;
- after processes with high defect rates;
- for goods receiving procedures with large batches of low cost items,
- destructive testing;
- inspection costs are high.

Sampling inappropriate:

- for small batches of high added value products;
- for safety/security products;
- for sensitive markets, e.g. initial supplies.

Pattern of sampling:

- Batch presented and random sample taken.
- If defective number less than or equal to a certain number then lot is accepted.
- Reject batches are 100% inspected.
  (i) Selecting sample size can be done using OC curves based upon agreed α, AWL and β LTPD. Alternatively a national standard (e.g. BS 6001) can be adopted based on batch size.

(ii) Acceptable quality levels are in the main market led through competition, supply contracts, market strategy, etc.
(iii) Sampling frequency is based upon batch availability and type of plan, e.g. single, multiple sequential, etc. Key is to define batch size.

$$AOQ = Pa\ PD\ \frac{(N - n)}{N}$$

Hence AOQ for Code B (at 4%) = 2.6
AOQ for Code J (at 4%) = 3.6
AOQL for Code J = 5.2
AOQL for Code B = 7.3

Though AOQ is higher with Code J the AOQL is lower. The AOQL with Code B occurs at a very high PD and therefore within production conditions plan B is preferred.

21. Characteristic reliability functions include:

- probability density functions;
- failure and reliability functions;
- age–specific failure rate;
- MTTF and MTBF.

Appropriate functions (exponential, Weibull, Erlang, etc.) can be used to model the reliability of process plant. Again costs can be considered in terms of:

- *Prevention*:
  - preventative maintenance programmes;
  - replanning and rescheduling;
  - maintenance staff and parts;
- *appraisal*:
  - monitoring systems and data collection;
  - inspection;
- *failure*:
  - breakdowns, scrap and lost production;
  - late delivery and market penalties;
  - imprecise planning.

Reliability functions are used to define maintenance periods and planning programmes to reduce failure costs through prevention. Prevention and appraisal costs can be managed and budgeted whereas failure costs are often uncontrolled.

Mean time between failure can be used to predict failure periods for plant which is maintained. Also important is the analysis of systems and relationships of components in series and parallel etc.

The Weibull plot give $\beta = 1$, $\eta = 2000$

$$\text{Hence } R(450) = \exp\left(-\left[\frac{t - t_0}{\eta - t_0}\right]^{\beta}\right)$$
$$= \exp\left(-\frac{450^1}{2000}\right)$$
$$= 0.8$$

Comments:

- $\beta = 1$, hence we are considering random failures which suggests the machines are within their working life.
- Duration 450 hours is low compared to the characteristic life of 2000 hours, hence reliability is high.

Reliability of systems in series:

$$R_{System} = R_1 \times R_2 \times R_3$$
$$= 0.8^3 = 0.512$$

This reduction in process reliability of 36% is offset by the improvement in machine availability of 40%. Alternatively consider 0.512 × 0.85 = 0.44 compared with 0.8 × 0.6 = 0.48.

Points to consider:

- Costs associated with reduction in reliability, time to repair disruption, downtimes, etc. What impact this would have upon quality/scrap and production planning would need to be considered.
- Improved availability would also need to be considered in terms of costs, e.g. tool rationalization, transportation costs, manning.

22. The candidates should be aware of the main features of sampling plans including:

- only a proportion of the batch is inspected, therefore reduced inspection costs;
- quantifiable probability that an acceptable batch will be rejected, i.e. producer's risk;
- similarly for the acceptance of an unacceptable quality level, i.e. consumer's risk;
- variety of plans can be selected on basis of cost, degree of control, etc. (can be single, double/multiple, sequential).

Sampling is appropriate where:

- testing is destructive,
- inspection costs are prohibitive;
- damage may occur due to handling etc.;
- before high cost processes;

- before defects become concealed,
- after operations with high defect rates;
- before change in responsibility, etc.

From text: AQL = 2%

n = 30 max. N = 250

α = 1–0.95 β = 0.05

From the chart for Pa = 0.95, PD = 2%

C	$\frac{PD \times n}{100}$	n
0	N/A	
1	0.36	18*
2	0.875	44
3	etc.	

As the sample size cannot exceed 30 the only plan available is:

C = 1 and n = 18

Then to plot the AOQ curve:

PD	$\frac{PD \times n}{100}$	Pa	AOQ $\frac{(PD \times Pa \times (N - n))}{N}$
1	0.18	0.985	0.16
2	0.36	0.95	0.32
3	0.54	0.9	0.45
4	0.72	0.84	0.56
5	0.90	0.77	0.64
6	1.08	0.71	0.71
7	1.26	0.65	0.76
8	1.44	0.58	0.78
9	1.62	0.54	0.81
10	1.80	0.47	0.79
11	1.98	0.42	0.77

Limit to the AOQL would be 0.81%.

23. (a) ISO 9000 covers a company's quality systems from material reception through to final test and despatch. The areas covered include:

- material reception and specification procedures;
- design and documentation (part 1) control;
- manufacturing procedures and control;
- final inspection and test procedures;
- specification of sampling plans;

- calibration systems;
- controls for non-conforming materials;
- identification of inspection standard;
- audit and review.

The implementation discussion should include:

- identification of appropriate part;
- identifying the quality objectives;
- identify responsible senior manager;
- planning to take into account all the company's activities;
- development of a quality policy and programme;
- document procedures and quality manual;
- review systems.

(b) The two forms of control charts for attribute data are proportion defective (p-chart) and number defective. Marks would be awarded for details of construction and applications.

24. Costs associated with the manufacturing process are as follows:

- *Failure*:
  - scrap and rework;
  - replanning and rescheduling;
  - warranty and delivery payments;
  - corrective servicing;
  - market share.
- *Prevention*:
  - inspection manpower and equipment;
  - supplier control and engineering standards;
  - process inspection and development;
  - training.

See Weibull plot $\beta = 1$ and $\eta = 3900$:

$$R(t) = \exp(-[\frac{t - t_0}{\eta - t_0}]^{\beta})$$

hence at 1000 hours:

$$R(1000) = \exp[-(\frac{1000}{3900})]^1]$$
$$= 0.77$$

The reliability of the system R_S is given by:

$$R_S = R_R[1-(1-R_M)_2]+(1-R_R)[1-(1-R_M{}^2)^2]$$
$$= 0.9[1-(1-0.77)^2]+0.1[1=(1-0.77^2)^2]$$
$$= 0.936$$

Selected Bibliography

The following books have been selected as recommended further reading for each of the chapters.

Chapter 1 – Introduction

- Cullen, J. and Hollingum, J. (1987) *Implementing Total Quality Management*, IFS.
- Feigenbaum, A.V. (1983) *Total Quality Control*, McGraw-Hill.
- Deming, W.E. (1986) *Out of the Crisis*, Cambridge University Press.
- Pascale, R. (1991) *Managing on the Edge*, Penguin.

Chapter 2 – Quality systems and ISO 9000

- British Standards Institution (1994) *BS EN ISO 9000 Quality Systems Model for Quality Assurance*, BSI.
- Stebbings, L. (1994) *Quality Assurance: The Route to Efficiency and Competitiveness*, Chapman & Hall
- Fox, M.J. (1995) *Quality Assurance Management*, Chapman & Hall.
- McRobb, R. (1989) *Writing Quality Manuals*, IFS.
- Arter, D.R. (1989) *Quality Audits for Improved Performance*, ASQC Quality Press.
- Ford (1987) *Manufacturing Guideline – Q101 SQA System Survey*, Ford Motor Company.

Chapter 3 – Quality costs and performance measurement

- British Standards Institution (1992) *BS 6143 Part 1: Process Cost Model*, BSI.
- British Standards Institution (1990) *BS 6143 Part 2: Prevention Appraisal and Failure Model*, BSI.
- Dale, B.G. and Plunkett, J.J. (1995) *Quality Costing*, Chapman & Hall.
- Zairi, M. (1994) *Measuring Performance for Business Results*, Chapman & Hall.

Chapter 4 – Motivation for quality

- Bentley, T. (1994) *Facilitation*, McGraw-Hill.
- Chopin, J. (1991) *Quality Through People*, IFS.
- Crosby, P. (1979) *Quality is Free*, McGraw-Hill.
- Labovitz, G., Chang, Y.S. and Rosansky, V. (1993) *Making Quality Work*, Harper Business.
- Peters, T. and Waterman, R.H. (1982) *In Search of Excellence*, Harper & Row

Chapter 5 – Total quality management

- Haavind, R. (1992) *The Road to the Baldridge Award*, Butterworth-Heinemann.
- Walton, M. (1990) *The Deming Management Method*, ASQC Quality Press.
- Ernst & Young (1990) *Total Quality – An Executive's Guide for the 1990s*, Ernst & Younge.
- Howe (1993) *Quality on Trial*, McGraw-Hill Book Company Europe.
- Hammer, M. and Champy, J. (1993) *Re-engineering the Corporation. A Manifesto for Business Revolution*, Harper Business.
- Dean, J.W. and Evans, J.R. (1994) *Total Quality – Management Organization and Strategy*, West.
- Oakland, J.S. (1991) *Total Quality Management*, Heinemann.

Chapter 6 – Acceptance sampling

- British Standards Institution (1991) *BS 6001 Acceptance Sampling by Attributes*, BSI.
- Dodge, H.F. and Romig, H.G. (1941) *Single Sampling and Double Sampling Inspection Tables*, Bell System Technical Journal.

Chapter 7 – Statistical process control

- Ishikawa, K. (1991) *Introduction to Quality Control*, Chapman & Hall.
- Oakland, J.S. (1986) *Statistical Process Control – A Practical Guide*, Heinemann.
- Besterfield, D.H. (1994) *Quality Control*, Prentice-Hall International.
- Owen, M. (1989) *SPC and Continuous Improvement*, IFS.
- Bissell, D. (1993) *Statistical Methods for SPC and TQM*, Chapman & Hall.

- Ford (1990) *Statistical Process Control*, Ford Motor Company.
- Murdoch, J. and Barnes, J.A. (1975) *Statistical Tables for Science, Engineering, Management and Business Studies*, Macmillan.
- Montgomery, D.C. (1991) *Introduction to Statistical Quality Control*, John Wiley and Sons.

Chapter 8 – Problem-solving tools

- Bergman, B. and Klefsjo, B. (1994) *Quality – From Customer Needs to Customer Satisfaction*, McGraw-Hill.
- Evans, J.R. and Lindsay, W.M. (1993) *The Management and Control of Quality*, West.
- Mizuno, S. (1988) *Management for Quality Improvement: The 7 New Q.C. Tools*, Productivity Press.
- Gitlow, H.S., Gitlow, S.J., Oppenheim, A. and Oppenheim, R. (1989) *Tools and Methods for the Improvement of Quality*, Irwin.

Chapter 9 – Reliability management

- Juran, J.M. and Gryna, F.M. (1980) *Quality Planning and Analysis*, McGraw-Hill.
- Henly, E.J. and Kumamoto, H. (1981) *Reliability Engineering and Risk Assessment*, Prentice-Hall.

Chapter 10 – Advanced quality planning

- Sullivan, L.P. (1986) *Quality Function Deployment*, Quality Progress.
- Bendell, A. *et al.* (1989) *Taguchi Methods – Applications in World Industry*, IFS.
- Akao, Y. (1990) *Quality Function Deployment*, Productivity Press.
- Ross, P.J. (1988) *Taguchi Techniques for Quality Engineering*, McGraw-Hill.

Index

Lightning Source UK Ltd.
Milton Keynes UK
UKOW07f1250080215

245862UK00007B/103/P

9 780412 626906